AF616375

Selected Titles in This Series

159 **S. Ya. Khavinson,** Best approximation by linear superpositions (approximate nomography), 1997

158 **Hideki Omori,** Infinite-dimensional Lie groups, 1997

157 **V. B. Kolmanovskiĭ and L. E. Shaĭkhet,** Control of systems with aftereffect, 1996

156 **V. N. Shevchenko,** Qualitative topics in integer linear programming, 1996

155 **Yu. Safarov and D. Vassiliev,** The asymptotic distribution of eigenvalues of partial differential operators, 1996

154 **V. V. Prasolov and A. B. Sossinsky,** Knots, links, braids and 3-manifolds. An introduction to the new invariants in low-dimensional topology, 1996

153 **S. Kh. Aranson, G. R. Belitsky, and E. V. Zhuzhoma,** Introduction to the qualitative theory of dynamical systems on surfaces, 1996

152 **R. S. Ismagilov,** Representations of infinite-dimensional groups, 1996

151 **S. Yu. Slavyanov,** Asymptotic solutions of the one-dimensional Schrödinger equation, 1996

150 **B. Ya. Levin,** Lectures on entire functions, 1996

149 **Takashi Sakai,** Riemannian geometry, 1996

148 **Vladimir I. Piterbarg,** Asymptotic methods in the theory of Gaussian processes and fields, 1996

147 **S. G. Gindikin and L. R. Volevich,** Mixed problem for partial differential equations with quasihomogeneous principal part, 1996

146 **L. Ya. Adrianova,** Introduction to linear systems of differential equations, 1995

145 **A. N. Andrianov and V. G. Zhuravlev,** Modular forms and Hecke operators, 1995

144 **O. V. Troshkin,** Nontraditional methods in mathematical hydrodynamics, 1995

143 **V. A. Malyshev and R. A. Minlos,** Linear infinite-particle operators, 1995

142 **N. V. Krylov,** Introduction to the theory of diffusion processes, 1995

141 **A. A. Davydov,** Qualitative theory of control systems, 1994

140 **Aizik I. Volpert, Vitaly A. Volpert, and Vladimir A. Volpert,** Traveling wave solutions of parabolic systems, 1994

139 **I. V. Skrypnik,** Methods for analysis of nonlinear elliptic boundary value problems, 1994

138 **Yu. P. Razmyslov,** Identities of algebras and their representations, 1994

137 **F. I. Karpelevich and A. Ya. Kreinin,** Heavy traffic limits for multiphase queues, 1994

136 **Masayoshi Miyanishi,** Algebraic geometry, 1994

135 **Masaru Takeuchi,** Modern spherical functions, 1994

134 **V. V. Prasolov,** Problems and theorems in linear algebra, 1994

133 **P. I. Naumkin and I. A. Shishmarev,** Nonlinear nonlocal equations in the theory of waves, 1994

132 **Hajime Urakawa,** Calculus of variations and harmonic maps, 1993

131 **V. V. Sharko,** Functions on manifolds: Algebraic and topological aspects, 1993

130 **V. V. Vershinin,** Cobordisms and spectral sequences, 1993

129 **Mitsuo Morimoto,** An introduction to Sato's hyperfunctions, 1993

128 **V. P. Orevkov,** Complexity of proofs and their transformations in axiomatic theories, 1993

127 **F. L. Zak,** Tangents and secants of algebraic varieties, 1993

126 **M. L. Agranovskiĭ,** Invariant function spaces on homogeneous manifolds of Lie groups and applications, 1993

125 **Masayoshi Nagata,** Theory of commutative fields, 1993

124 **Masahisa Adachi,** Embeddings and immersions, 1993

(Continued in the back of this publication)

Best Approximation by Linear Superpositions (Approximate Nomography)

Translations of

MATHEMATICAL MONOGRAPHS

Volume 159

Best Approximation by Linear Superpositions (Approximate Nomography)

S. Ya. Khavinson

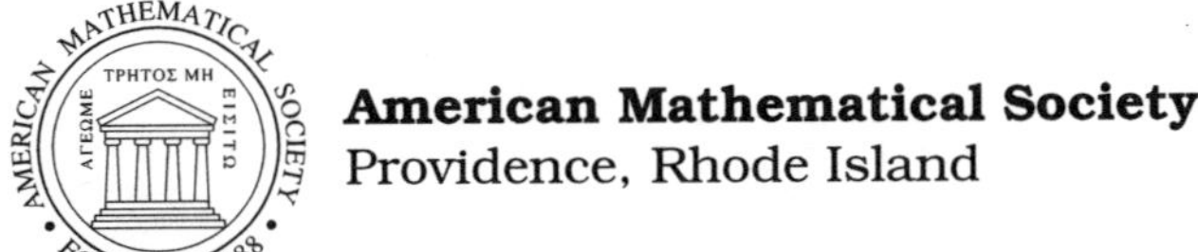
American Mathematical Society
Providence, Rhode Island

С. Я. Хавинсон

НАИЛУЧШЕЕ ПРИБЛИЖЕНИЕ
ЛИНЕЙНЫМИ СУПЕРПОЗИЦИЯМИ
(АППРОКСИМАТИВНАЯ НОМОГРАФИЯ)

Translated by D. Khavinson from an original Russian manuscript

1991 *Mathematics Subject Classification.* Primary 41–XX; Secondary 41A50, 41A52.

Abstract. Problems concerning the approximation of functions of several variables by superpositions of functions of fewer variables originated from Kolmogorov's celebrated theorem. This book presents the current state of the theory of approximation by superpositions in various function spaces.

The book is intended for research mathematicians and graduate students working in approximation theory, constructive function theory, and functional analysis.

Library of Congress Cataloging-in-Publication Data

Khavinson, S. ĪA.

Best approximation by linear superpositions (approximate nomography) / S. Ya. Khavinson ; [translated by D. Khavinson from an original Russian manuscript].

p. cm.—(Translations of mathematical monographs ; v. 159)

Includes bibliographical references (p. –).

ISBN 0-8218-0422-7 (alk. paper)

1. Approximation theory. 2. Functions of several real variables. 3. Functional analysis. I. Title. II. Series.

QA221.K467 1996

511′.4—dc20

96-36520
CIP

∞ The paper used in this book is acid-free and falls within the guidelines
established to ensure permanence and durability.

10 9 8 7 6 5 4 3 2 1   02 01 00 99 98 97

Contents

Introduction 1

Chapter 1. Discussing Kolmogorov's Theorem 5
- §1. A. N. Kolmogorov's theorem 5
- §2. Duality in problems concerning representations by superpositions 15
- §3. Separation of points and measures 22
- §4. Constructing function families separating Borel measures 29
- §5. Dimension and the number of terms in the Kolmogorov representation 32
- §6. Other types of separation 34
- §7. Study of the original notion of separation (continued) 38
- §8. A counterexample 40
- §9. Measure of compact sets on which all continuous functions are representable by sums of superpositions 50

Chapter 2. Approximation of Functions of Two Variables by Sums $\varphi(x) + \psi(y)$ 53
- §1. Raising the questions. Lightning bolts 53
- §2. Closedness of the subspace D 58
- §3. Proximinality 68
- §4. Annihilator of sums of superpositions. When is the subspace of sums of superpositions everywhere dense? 73
- §5. Relation to the theory of functional equations 89
- §6. Chebyshev-like problems for the best approximation of a function of two variables by sums $\varphi(x) + \psi(y)$ 92
- §7. The levelling algorithm 112

Chapter 3. Problems of Approximation by Linear Superpositions 127
- §1. Properties of the subspace of linear superpositions and its annihilator 127
- §2. On the existence of best approximations 150
- §3. Effective construction of best approximations 157

References 169

Introduction

In 1957, while studying Hilbert's famous thirteenth problem, A. N. Kolmogorov [**86**] obtained the following remarkable result. Set $I = [0,1]$, $\mathbb{R} = (-\infty, +\infty)$.

KOLMOGOROV'S THEOREM. *There exist increasing functions* $\varphi_{pq}(x) \in C(I)$, $p = 1, \dots, n$, $q = 1, \dots, 2n+1$, *such that an arbitrary given function*

$$f(x_1, \dots, x_n) \in C(I^n)$$

can be represented in the form

$$f(x_1, \dots, x_n) = \sum_{q=1}^{2n+1} g_q \left(\sum_{1}^{n} \varphi_{pk}(x_p) \right), \tag{0.1}$$

where $g_q \in C(\mathbb{R})$ *depend on* f.

Thus, every continuous function of n variables can be represented by superpositions of continuous functions of one variable and the simplest function of two variables, namely, the sum of the variables. Earlier, Arnold [**4**], [**5**] showed that $f \in C(I^3)$ can be represented by superpositions of continuous functions of two variables, while Kolmovorov himself [**86**] proved that $f \in C(I^n)$ is always representable as a superposition of continuous functions of three variables. Arnold's theorem has given a negative answer to Hilbert's conjecture. (About Hilbert's problems, we refer the reader to [**72**], [**143–144**], [**145**], [**125**], [**95**], [**97–98**].)

Later on, Kolmogorov's theorem was extended and improved by a great many authors. (Of course, continuity is an essential requirement for representability by superpositions to become a deep and important result. When it is removed, the possibility of such representation is almost trivial. Let, say, $f(x_1, x_2, x_3)$ be an arbitrary function on I^3, and let $x_1 = \Phi_1(t)$, $x_2 = \Phi_2(t)$ map I on I^2. If $t = \Phi(x_1, x_2)$ is any single-valued branch of the inverse map, it is necessarily discontinuous. Consider a function $g(t, \tau) = f(\Phi_1(t), \Phi_2(t), \tau)$ on I^2. Then

$$f(x_1, x_2, x_3) = g[\Phi(x_1, x_2), x_3],$$

and we obtain representation of an arbitrary function of three variables by a superposition of functions of two variables, although Φ is a discontinuous function here.)

Besides problems of finding precise expressions for a given function in terms of a combination of functions of fewer variables, it is natural to consider problems concerning the best approximation of functions of several variables by combinations (of a special type) of functions of fewer variables. Back in 1938, Denisyuk [**37**] considered the following approximation problem: for $f(x,y) \in L^2(I^2)$ find

$$\inf_{\substack{\varphi\in L^2(I)\\ \psi\in L^2(I)}} \iint_{I^2} |f(x,y)-\varphi(x)-\psi(y)|^2\,dxdy.$$

The space L^2 being a Hilbert space, we can easily compute the best approximation

$$\varphi_0(x)+\psi_0(y)=\int_0^1 f(x,y)dy+\int_0^1 f(x,y)dx-\iint_{I^2} f(x,y)dxdy. \tag{0.2}$$

A similar problem in the space $C(I^2)$, first studied by Diliberto and Straus [**38**], is much more delicate. Independently, the latter problem was studied by Aumann [**9–13**]. In Ofman's paper [**113**], presenting results of Kolmogorov, Smolyak, Arnold, and Ofman himself, this problem was also studied independently of [**38**], and I^2 was replaced by an arbitrary set in the plane. In their paper, Diliberto and Straus initiated study of a natural procedure, a "levelling algorithm", for the construction of the best approximation $\varphi^\star(x)+\psi^\star(y)$ of the function $f(x,y)$. The same process, "by analogy", was suggested in [**38**] for the case where the number of variables is greater than two. However, Aumann [**13**] showed that already for the case of three variables, this algorithm does not lead to the corresponding best approximation and the calculation of its values. (Unfortunately, this work has not been widely known.)

Golomb's paper [**64**] can be viewed as the starting point of a systematic study of approximation of functions of several variables by various combinations (including nonlinear ones) of functions of fewer variables in the L^2-metric and in C. He has also described a more abstract version of the algorithm of Diliberto and Straus and an extension of that version to the case with a greater number of terms. However, in [**13**] it was also shown that this abstract version of the Diliberto–Straus algorithm cannot be extended to the case with more than two variables.

Aumann suggested calling such approximation problems "approximate nomography". As is known, for a functional dependence to be nomographable, i.e., representable by a special system of graphic images (cf., e.g., [**45a**]), it is necessary that it admit a representation in terms of superpositions of a special kind $(g(\varphi(x)+\psi(y)))$. According to this viewpoint, consideration of precise representations (similar to (0.1)) is "nomography", while approximation by certain superpositions is "approximate nomography". In [**37**] it is also noted that the problem appeared due to the necessity of being able to approximate nomographically. However, one has to keep in mind that Hilbert himself, while stating in his thirteenth problem ([**72**]) the question on superpositions, associated it directly with nomography, which had just then appeared in the fundamental treatise of M. d'Ocagne, "Traité de Nomographie", Paris, 1899.

Also, note that in [**38**] the authors point out that their problem appeared in the framework of studies carried out by the Rand Corporation (and quite likely was associated with the possibility of the most economical storage of information in the memory (of a quite modest size) of contemporary computers).

The above results on representation and approximation of functions of several variables by combinations of functions of fewer variables can be included (for the C-metric) into the following scheme. Let compact spaces X, $\{X_i\}$ and continuous maps $\Phi_i : X \to X_i$, $i = 1, \dots, N$, be given; we wish to study properties of the subspace $D \subset C(X)$ that consists of the following sums of superpositions:

$$D = \left\{ \sum_{i=1}^{N} g_i \circ \Phi_i(x),\ g_i \in C\left(X_i\right), i = 1, \dots, N \right\}. \tag{0.3}$$

We are interested in the following properties of the subspace D: conditions when D coincides with $C(X)$, i.e., $D = C(X)$; density of D in $C(X)$, closeness of D in $C(X)$, proximinality of D, etc. The question is posed as follows: what requirements must one impose on the configuration $(X, \{X_i\}, \{\Phi_i\})$ in order that D possess one of the above properties? The problem can also be interpreted as the study of properties of the sum of closed subalgebras in the algebra $C(X)$. For the study of equality $D = C(X)$, the most crucial step after Kolmogorov's theorem was an approach based on duality introduced by Sternfeld in [**133–136**]. It turned out that the conditions ensuring the equality $D = C(X)$ are realized as properties of the system $\Phi_1, \dots, \Phi_N$ to separate points or regular Borel measures on Y in some quite strong sense. Starting out from this observation, Sternfeld, in a difficult paper [**135**], showed that the number of terms in the Kolmogorov theorem (0.1) cannot be made smaller than $2n + 1$. (Some preliminary results in those directions can be found in [**132**], [**133**].)

Besides the above questions concerning properties of D, it is of interest to study the annihilator $D^{\perp}$ in $C(X)^{*}$, approximation of an individual function $f \in C(X)$ by functions in D: duality, existence and characterization of the best approximation of f, its uniqueness (which almost always fails in such problems in the C-metric), and algorithms for constructing the best approximation. Similar questions arise for more general subspaces

$$D = \left\{ \sum_{1}^{N} h^{i}(x) \left[g_i \circ \Phi_i(x)\right],\ g_i \in C\left(X_i\right) \right\}, \tag{0.4}$$

where the $h^i(x)$ are given functions on X. Functions like those in (0.4) are called linear superpositions (A. G. Vitushkin). The most important example of linear superpositions is the following. Let (j) be a set of indices selected from $\bar{N} = \{1, \dots, N)$; $(\tilde{j})$ denotes a complementary set of indices to (j) (in $\bar{N}$). Set $Y_{(j)} = X_{i_1} \times \cdots \times X_{i_j}$, where $(j) = (i_1, \dots, i_j)$, and $Y = X_1 \times \cdots \times X_N$. Let $(j_1), \dots, (j_m)$ be given index sets such that none of them completely includes another (non-empty intersections are allowed, though). Let H^j be a finite-dimensional subspace in $C(Y_{(\tilde{j})})$, and

$$D = \sum_{1}^{m} H^{j} \otimes C(Y_{(\tilde{j})}). \tag{0.5}$$

In particular, for H^j one can take trigonometric polynomials (of several variables), provided that Y is a finite-dimensional Euclidean space. In that case, functions in D are sometimes referred to as generalized polynomials, or quasi-polynomials.

The list of problems discussed above (with some obvious modifications) is equally worth studying for other function spaces, as well. However, I have chosen to confine myself exclusively to the spaces C and ℓ^{∞}. Only in §3 of the last chapter do I present a result regarding the space L^2, one which provides a far-reaching extension of the solution (0.2).

The problems that we are dealing with in this monograph have attracted the attention of a large number of investigators, and the theory continues to develop actively. On one side, theory of approximation by linear superpositions provides a rich soil for models and development of general methods in Banach spaces, and contains a number of difficult problems. On the other side, it itself contains many possibilities for various applications (including numerical computations). Some developments of the theory were presented in the 1985 monograph by Cheney and Light [**94**], where approximation problems for a subspace D of (0.5)-type in two variables ($N = m = 2$) were considered. The book [**94**] also contains an extensive bibliography. The subject is, however, much larger than presented there, and, in particular, some important new results were obtained after [**94**] appeared. This allowed me to select the material for the present book in such a way that the overlaps with [**94**] are minimal. Formally, it overlaps with [**94**] only in §7 of Chapter 2, where we study the algorithm of Diliberto and Straus. Though we only consider the case of the uniform metric and the subspace of sums (0.3) ($N = 2$), we go into much greater detail than [**94**] does. (In [**94**] the authors also considered not only subspaces (0.5) for $N = m = 2$, but different metrics as well.)

In principle, investigations of superpositions of differentiable functions initiated by A.G. Vitushkin and later on studied in his joint work with G.M. Khenkin and by numerous other investigators, form an important part of the theory of superpositions. We have not pursued these developments here (let us cite the surveys [**144**], [**145**], [**137**], [**95**], [**98**]).

Throughout the book, all propositions (theorems, lemmas, corollaries, etc.) are enumerated with the number of the section in which they appear and the number of the proposition itself in the section. Whenever we refer to a result from a different chapter, we also specify the chapter number. In some cases, for the reader's convenience, instead of referring to the preceding exposition, we reproduce the basic setting under investigation once more.

CHAPTER 1

Discussing Kolmogorov's Theorem

§1. A. N. Kolmogorov's theorem

The statement. The statement of Kolmogorov's Theorem given in the Introduction has been improved by many authors. G. Lorentz [**35**] observed that one could choose the functions g_k in (0.1) to be the same. Sprecher [**123**] showed that the functions φ_{pq} in (0.1) can be replaced by $\lambda_p \varphi_q$ with appropriate constants λ_p.

So, one can talk about representing an arbitrary continuous function f on I^n (where $I = [0, 1]$) in the form

$$f(x_1, \dots, x_n) = \sum_{q=1}^{2n+1} g\left(\sum_{p=1}^{n} \lambda_p \varphi_q(x_p)\right). \tag{1.1}$$

T. Hedberg [**71**] and J.-P. Kahane found a proof of (1.1) based on the Baire category theorem.

Let us introduce some terminology:

Quasi-all points. Let X be a complete metric space. We shall say that a certain property holds for quasi-all points in X (holds quasi-everywhere in X) if it holds for all points of a set $U \subset X$ which is a countable intersection of open dense sets in X. (Therefore, the complement of U is a F_σ set of the first Baire category.)

Rationally independent numbers. Real numbers $\lambda_1, \dots, \lambda_n$ are called rationally independent if, for an arbitrary n-tuple of rational numbers $r_1, \dots, r_n$, not all of which are equal to zero, the following holds:

$$\lambda_1 r_1 + \cdots + \lambda_n r_n \neq 0.$$

Rationally independent numbers $\lambda_1, \dots, \lambda_n$ must all be different, and at least $n-1$ of them are irrational. Quasi-all vectors $\lambda = (\lambda_1, \dots, \lambda_n)$ of the space $\mathbb{R}^n$ have rationally independent coordinates. Indeed, any vector $\lambda = (\lambda_1, \dots, \lambda_n)$ with rationally dependent coordinates must belong to a hyperplane

$$r_1 \lambda_1 + \cdots + r_n \lambda_n = 0$$

with a normal vector $r = (r_1, \dots, r_n)$ with rational coordinates. Each such hyperplane is a closed, nowhere dense set in $X = \mathbb{R}^n$, and there are countably many of them.

If we introduce a norm in $\mathbb{R}^n$ (for our purposes we do it by setting $\|\lambda\| = \sum_1^n |\lambda_i|$), then quasi-all vectors on the surface Σ of the unit ball have rationally independent coordinates. The $(n-1)$-dimensional Lebesgue measure of the set Γ of those vectors equals the full measure of the surface Σ.

Kahane's reformulation. Let Φ be a set of continuous, non-decreasing functions φ on $I = [0,1]$ such that $\varphi(0) = 0$, $\varphi(1) = 1$. Since it is a closed subset in the space $C[0,1]$ of continuous functions on $[0,1]$ (with the usual norm), Φ is a complete metric space. It is not hard to see that quasi-all functions in Φ are in fact strictly increasing. Indeed, if $0 \le r_1 < r_2 \le 1$ are two rational numbers, then the set U_{r_1,r_2} of all functions $\varphi \in \Phi$ such that $\varphi(r_1) < \varphi(r_2)$ is open and everywhere dense in Φ. The intersection $U = \bigcap\limits_{r_1,r_2} U_{r_1,r_2}$ consists of all strictly-increasing functions, and is an intersection of a countable number of everywhere dense sets on Φ.

Set $\Phi^{2n+1} = \{(\varphi_1, \dots, \varphi_{2n+1}), \quad \text{where } \varphi_i \in \Phi,\ i = 1, \dots, 2n+1\}$. Φ^{2n+1} is a complete metric space as well (it is a closed subset of $[C(I)]^{2n+1}$).

In Kahane's reformulation, the Kolmogorov theorem has the following form:

THEOREM 1.1. *Let $\lambda_1 > 0, \dots, \lambda_n > 0$ be rationally independent numbers, $\sum\limits_1^n \lambda_p = 1$. For quasi-all n-tuples $(\varphi_1, \dots, \varphi_{2n+1}) \in \Phi^{2n+1}$ the following statement holds:*

For any $f \in C(I^n)$ there exists a function $g \in C(I)$ such that the representation (1.1) holds. This is true whenever $C(I^n)$ and $C(I)$ are simultaneously taken as spaces of real- or complex-valued functions.

Free interpolation in the choice of an outer function. As was observed in [**71**], [**74**], [**95**], one has a great deal of freedom in choosing an outer function g in the superposition (1.1). For example, g could be chosen among all restrictions to $[0,1]$ of functions $g(x) = \tilde{g}(e^{ix})$, $x \in [0, 2\pi]$, such that $\tilde{g}(z)$ is analytic in the disk $\{z : |z| < 1\}$ and continuous in the closed disk $\{z : |z| \le 1\}$.

We are going to associate this observation with a possibility of so-called free interpolation in choosing g. Let Y be a subspace of the space $C(K)$ of functions continuous on a compact set K; let $E \subset K$ be a closed subset. We say that Y interpolates freely on E if for each $H(x) \in C(E)$ there exists $g \in Y$ such that $g(x) = H(x)$ for $x \in E$. We say that Y interpolates freely on E with a constant c $(c \ge 1)$, if for each $H(x) \in C(E)$ there exists $g(x) \in Y$ such that

$$g(x) = H(x), \quad x \in E, \qquad \|g\|_{C(K)} \le c\|H\|_{C(E)}. \tag{1.2}$$

If Y is a closed subspace in $C(K)$ that interpolates freely on E, then free interpolation always occurs with a constant c.

An important example of free interpolation is given by the Rudin–Carleson theorem (see, e.g., [**55**]) concerning boundary values of continuous analytic functions. In relation to our situation, this theorem can be described as follows. Let Y consist of boundary values $g(x) = \tilde{g}(e^{2\pi i x})$, $0 \le x \le 1$, of functions $\tilde{g}(z)$ analytic in the unit disk $\{z : |z| \le 1\}$. Then Y interpolates freely with the constant $c = 1$ on every closed set $E \subset (0,1)$ that has measure zero.

THEOREM 1.2. *Let λ_j be as in Theorem 1.1, and let Y be a closed subspace in $C(I)$ that interpolates freely with a constant $c < \dfrac{2(n+1)}{2n+1}$ on a nowhere dense closed subset $E \subset (0,1)$ and on all sets obtained by adding finitely-many points from $[0,1]$ to E. Quasi-all $(2n+1)$-tuples $(\varphi_1, \dots, \varphi_{2n+1}) \in \Phi^{2n+1}$ have the following property: for any $f(x_1, \dots, x_n) \in C(I^n)$ and any $H \in C(E)$, there exists a function*

$g(x) \in Y$ such that (1.1) holds and

$$g(x) = H(x), \qquad x \in E. \tag{1.3}$$

COROLLARY 1.3. *In particular, the conclusion of Theorem 1.2 holds for $Y = \{g(x) : g(x) = \tilde{g}(e^{ix})\}$, where $\tilde{g}(z)$ is analytic in $\{z : |z| < 1\}$ and continuous in $\{z : |z| \le 1\}$.*

If $f(x_1, \dots, x_n)$ satisfies $f(0, \dots, 0) = f(1, \dots, 1)$, then one can choose $Y = \{g(x) : g(x) = \tilde{g}(e^{2\pi ix})\}$; the reason is that in the proof of Theorem 1.2, one must construct a function g that interpolates the value $f(0, \dots, 0)$ at the point $x = 0$ and the value $f(1, \dots, 1)$ at $x = 1$, while a function continuous on the circle must satisfy $g(0) = g(2\pi)$.

Lemmas. Consider a map $\lambda : \mathbb{R}^n \to \mathbb{R}$ defined by

$$a \in \mathbb{R}^n \mapsto \lambda(a) = \langle \lambda, a \rangle = \sum_1^n \lambda_k a_k, \tag{1.4}$$

where $\lambda = (\lambda_1, \dots, \lambda_n)$ is a vector whose coordinates satisfy the assumptions in Theorems 1.1 and 1.2, and $a = (a_1, \dots, a_n)$.

LEMMA 1.3. (1) *Let Q be the set of all points with rational coordinates in $\mathbb{R}^n$. The map λ is injective on Q.*

(2) *$\lambda(I^n) = I$, $\lambda(I_0^n) = I_0$ ($I = [0, 1]$, $I_0 = (0, 1)$).*

(3) *$\lambda^{-1}(E)$ is a nowhere dense closed set.*

PROOF. (1) follows from the rational independence of $\lambda_1, \dots, \lambda_n$. (2) is obvious. If we assume that $\lambda^{-1}(E)$ is dense somewhere, then it contains a ball S. But then $E \supset \lambda(S)$ must contain an interval, since a linear map λ is an open map.

LEMMA 1.4. *Let Λ be a restriction of λ onto I^n. Then:*

(1) *$\Lambda^{-1}(0) = (0, \dots, 0)$, $\Lambda^{-1}(1) = (1, \dots, 1)$.*

(2) *The set $U = \Lambda^{-1}(I_0 \backslash E)$ is open and everywhere dense in I^n.*

(3) *Let $U_{i_1 \dots i_k}$ ($i_1 < \dots < i_k$) be a set of points in U such that their coordinates with indices different from $i_1, \dots, i_k$ are fixed. For a point $a \in U_{i_1 \dots i_k}$, denote by a' a point in $\mathbb{R}^k$ whose coordinates coincide with free coordinates of the point a (i.e., coordinates with the indices $i_1, \dots, i_k$). The points a' form an open and everywhere dense set in I^k.*

PROOF. (1) follows since all $\lambda_p > 0$ and $\sum_1^n \lambda_p = 1$. (2) follows from statement (3) of the preceding lemma. Let us focus on (3). Let $a_{i_1}, \dots, a_{i_k}$ be those coordinates of a that are free to vary, and b_j, $j \ne i_1, \dots, i_k$, be fixed coordinates of a. Then,

$$\Lambda(a) = \langle \lambda, a \rangle = \sum_{\ell=1}^k \lambda_{i_\ell} a_{i_\ell} + \sum_{j \ne i_1, \dots, i_k} \lambda_j b_j. \tag{1.5}$$

It is clear that (1.5) defines an affine map $\tilde{\Lambda}$ of $I_{i_1} \times \dots \times I_{i_k}$ into I:

$$\tilde{\Lambda} : a' = (a_{i_1}, \dots, a_{i_k}) \to \Lambda(a).$$

As in (3) of the preceding lemma, we can show that $\tilde{\Lambda}^{-1}(E)$ is closed and nowhere dense. Hence, $U_{i_1 \dots i_k} = \tilde{\Lambda}^{-1}(I_0 \backslash E)$ is open and everywhere dense in I^k.

LEMMA 1.5. *Let* $0 = a_1 < a_2 < \cdots < a_m = 1$, $m > n+2$. *Let a number* $\varepsilon > 0$ *be chosen so that* $2\varepsilon < \min\limits_{1 \le i \le m-1} (a_{i+1} - a_i)$. *Set* $S_i = (a_i - \varepsilon, a_i + \varepsilon)$ *for* $i = 2, \dots, m-1$, *and* $S_1 = \{0\}$, $S_m = \{1\}$. *There exist intervals* $\sigma_i \subset S_i$, $i = 2, \dots, m-1$, *so that (for the sake of notational symmetry we also define* $\sigma_1 = S_1 = \{0\}$, $\sigma_m = S_m = \{1\}$*) the Cartesian product of any* n *of the intervals* $\sigma_1, \dots, \sigma_m$ *belongs to the set* $U = \Lambda^{-1}(I_0 \backslash E)$. *Moreover, at different places in those products we can take the same intervals, and the order of the terms in the product need not coincide with the order determined by the indices of those intervals.*

PROOF. Consider all possible permutations with repetitions of n numbers taken from the numbers $1, \dots, m$. Let $(i_1, \dots, i_n)$ be such a permutation. The parallelepiped $S_{i_1} \times \cdots \times S_{i_n} \subset I^n$ contains a parallelepiped $\sigma^1_{i_1} \times \cdots \times \sigma^1_{i_n}$, $\sigma^1_{i_k} \subset S_{i_k}$, that is entirely contained in U. If there are no ones or zeros among the numbers $i_1, \dots, i_n$ this follows from (2) of Lemma 1.4, while if there are some ones (or zeros) it follows from (3) of Lemma 1.3. The upper index denotes the number of the step of the argument. Let us now take another permutation $(j_1, \dots, j_n)$ with repetitions out of the indices $1, \dots, m$. Taking the product $S_{j_1} \times \cdots \times S_{j_n}$, replace in it factors with the indices that have already appeared in the preceding set $(i_1, \dots, i_n)$ by σ^1. For example, if j_ℓ has already been chosen among $i_1, \dots, i_n$, then we replace S_{j_ℓ} by $\sigma^1_{j_\ell}$. Once again, according to (2), (3) of Lemma 1.4 there exists a parallelepiped $\sigma^2_{j_1} \times \cdots \times \sigma^2_{j_n}$ that entirely belongs to U and, moreover, $\sigma_{j_\ell} \subset S_{j_\ell}$ $(\ell = 1, \dots, n)$. For those numbers from the first collection that are not included in the second, we simply change the upper index from 1 to 2 for corresponding intervals σ. Take the next set of indices $(k_1, \dots, k_n)$ and consider the product $S_{k_1} \times \cdots \times S_{k_n}$. For those indices that have already been chosen in the first or second collection, replace the factors in this product by smaller intervals σ^2_k. Once again, there exist smaller intervals $\sigma^3_{\ell_k} \subset S_{\ell_k}$ so that their Cartesian product is contained in U. For those indices from the first two collections that do not belong to the third collection, now simply replace for the corresponding intervals σ the upper index 2 by 3 and pass on to the next set of indices. After $N = m^n$ steps, when we have considered all possible permutations of n indices, we obtain intervals $\sigma^N_1, \dots, \sigma^N_m$ satisfying the statement of the lemma (and we can remove the upper index).

The Main Lemma. The following lemma plays the major role in the proof of Theorem 1.2. It is a natural extension to a more complicated situation of the main step in the proof of the Kahane–Hedberg Theorem 1.1.

LEMMA 1.6. *Let the numbers* $\lambda_1, \dots, \lambda_n$, *a set* E, *a subspace* $Y \subset C[0,1]$, *and a constant* c *satisfy the assumptions of Theorem* 1.2. *Choose a number* ε_0, $0 < \varepsilon_0 < \dfrac{1}{2(n+1)}$, $(1-\varepsilon_0)\,c < 1$. *According to the conditions imposed on* c, *we have*

$$\left(1 - \frac{1}{2(n+1)}\right) c = \frac{2n+1}{2(n+1)}, \qquad c < 1.$$

So, if ε_0 *is less than but sufficiently close to* $\frac{1}{2(n+1)}$, *we shall have* $(1-\varepsilon_0)\,c < 1$. *For all* $F(x_1, \dots, x_n) \in C(I^n)$ *and all* $H(x) \in C(E)$, *denote by* $\Omega(F, H)$ *the set of all collections* $(\varphi_1, \dots, \varphi_{2n+1}) \in \Phi^{2n+1}$ *for which there exists a function* $h \in Y$ *with the following properties*:

$$\left\| F(x_1, \dots, x_n) - \sum_{q=1}^{2n+1} h\left(\sum_{p=1}^{n} \lambda_p \varphi_q(x_p)\right) \right\| < (1 - \varepsilon_0)\, c \max(\|F\| \|H\|), \tag{1.6}$$

$$h(x) = 2\varepsilon_0 H(x), \qquad x \in E, \quad \|h\| \le 2\varepsilon_0 c \max(\|F\|, \|H\|). \tag{1.7}$$

Then $\Omega(F, H)$ *is an open everywhere dense set in* Φ^{2n+1}.

PROOF. The fact that $\Omega(F, H)$ is an open set is obvious, since when conditions (1.6) and (1.7) hold for some collection $(\varphi_1^0, \dots, \varphi_{2n+1}^0) \in \Phi^{2n+1}$ with a function $h \in Y$, they will also hold for all collections in Φ^{2n+1} sufficiently close to $(\varphi_1^0, \dots, \varphi_{2n+1}^0)$. Let us check that $\Omega(F, H)$ is everywhere dense in Φ^{2n+1}. Let $\mathcal{P}$ be an arbitrary open set in Φ^{2n+1}. We have to show that the intersection of $\Omega(F, H)$ and $\mathcal{P}$ is non-empty. Denote by $\delta = \delta(\varepsilon, \mathcal{P}, F, H)$ a small positive number to be chosen later and consider $(2n+1)$ intervals

$$\begin{gathered} I_q = \{I_q(j)\} = \{[q\delta + (2n+1)j\delta,\ q\delta + (2n+1)j\delta + 2n\delta]\}, \\ j = 0, \pm 1, \dots; \qquad q = 1, \dots, 2n+1. \end{gathered} \tag{1.8}$$

Note the following:

1°. For a given q, the intervals $I_q(j)$ are disjoint and separated by intervals of length δ.

2°. Every point in $I = [0, 1]$ is covered by intervals of all series I_q, except perhaps for one value of q.

Consider now the series of cubes P_q of dimension n:

$$P_q = \{P_q(j_1, \dots, j_n) = I_q(j_1) \times I_q(j_2) \times \dots \times I_q(j_n)\}, \tag{1.9}$$

where $(j_1, \dots, j_n)$ is an arbitrary permutation with repetitions of indices j.

According to 2°, each point of the unit cube I^n can fail to be covered by cubes P_q out of no more than n series, and hence is covered by cubes P_q of at least $n+1$ series. Let Δ be a set of collections $(\varphi_1, \dots, \varphi_{2n+1}) \in \Phi^{2n+1}$ such that φ_q are constants on intervals of the system I_q and linear on the intervals in-between $I_q(j)$ and $I_q(j+1)$. Now choose $\delta > 0$ so that the following hold:

(a) The oscillation of F on any of the cubes P_q does not exceed $\varepsilon\|F\|$ (in the case when F is a complex-valued function, by the oscillation of F on P_q we understand the diameter of the set $F(P_q)$).

(b) $\mathcal{P} \cap \Delta \neq \emptyset$.

Clearly, (a) and (b) are satisfied if we choose $\delta > 0$ sufficiently small. Let $(\varphi_1^0, \dots, \varphi_{2n+1}^0) \in \Phi^{2n+1}$ belong to $\mathcal{P} \cap \Delta$. Denote a constant value of φ_q^0 on the interval $I_q(j)$ by $\varphi_q^0(j)$. We can assume that the values of all functions φ_q^0 are different, i.e., $\varphi_q(j_1) \neq \varphi_r(j_2)$, provided that $q \neq r$, or that $q = r$ but $j_1 \neq j_2$. This holds for all intervals, except for those containing 0 or 1 (if the series I_q contains such intervals), because $\varphi_q^0(0) = 0$, $\varphi_q^0(1) = 1$ always. Apply Lemma 1.5, taking for the numbers a_j the values $\varphi_q^0(j)$. Accordingly, we can replace the values $\varphi_q^0(j)$ on $I_q(j)$ by rational values $r_q(j)$ that are arbitrarily close to $\varphi_q^0(j)$ (and hence are different) and such that the point $(\varphi_q(j_1), \dots, \varphi_q(j_n))$ for each permutation $(j_1, \dots, j_n)$ belongs to the set $U = \Lambda^{-1}(I_0 \backslash E)$ (Lemmas 1.4 and 1.5). Interpolating linearly between the intervals $I_q(j)$, we obtain functions φ_q such that $(\varphi_1, \ldots, \varphi_{2n+1}) \in \mathcal{P} \cap \Delta$. Now consider the functions

$$\psi_q(x_1, \ldots, x_n) = \sum_{p=1}^{n} \lambda_p \varphi_p(x_p) = \Lambda(\varphi_q(x_1), \ldots, \varphi_q(x_n)). \tag{1.10}$$

On each of the cubes $P_q(j_1, \ldots, j_n)$ the functions ψ_q take constant values

$$\Lambda(\varphi_q(j_1), \ldots, \varphi_q(j_n)) \in I_0 \backslash E,$$

provided that the cube $P_q(j_1, \ldots, j_n)$ does not contain $(0, \ldots, 0)$ and $(1, \ldots, 1)$. If $(0, \ldots, 0)$ or $(1, \ldots, 1)$ belongs to that cube the values of ψ_q will be 0 or 1, accordingly.

Finally, consider a map $\mathcal{X}(q; j_1, \ldots, j_n)$ of all sets $(q; j_1, \ldots, j_n)$ into I defined as follows:

$$\mathcal{X}(q; j_1, \ldots, j_n) = \psi_q(x_1, \ldots, x_n) = \Lambda(\varphi_q(x_1), \ldots, \varphi_q(x_n)),$$
$$(x_1, \ldots, x_n) \in I_q(j_1, \ldots, j_n).$$

If $P_q(j_1, \ldots, j_n) \ni (0, \ldots, 0)$, then $\mathcal{X}(q; j_1, \ldots, j_n) = 0$. If $P_q(j_1, \ldots, j_n) \ni (1, \ldots, 1)$, then $\mathcal{X}(q; j_1, \ldots, j_n) = 1$. In both cases, the n-tuples $(j_1, \ldots, j_n)$ clearly contain the same numbers. $\mathcal{X}$ maps injectively all other selections $(q; j_1, \ldots, j_n)$ into $I_0 \backslash E$ (because the values $\varphi_q(j)$ are different rational numbers, while the numbers $\lambda_1, \ldots, \lambda_n$ are rationally independent).

Taking a cube $P_q(j_1, \ldots, j_n)$, define on it the value $M(q; j_1, \ldots, j_n)$ of the function $F(x_1, \ldots, x_n)$ as follows: if $P_q(j_1, \ldots, j_n)$ does not contain $(0, \ldots, 0)$ and $(1, \ldots, 1)$, then $M(q; j_1, \ldots, j_n)$ can be any value of $F(x_1, \ldots, x_n)$ on that cube. If $P_q(j_1, \ldots, j_n)$ contains $(0, \ldots, 0)$, then $M(q; j_1, \ldots, j_n) = F(0, \ldots, 0)$. If $P_q(j_1, \ldots, j_n)$ contains $(1, \ldots, 1)$, then $M(q; j_1, \ldots, j_n) = F(1, \ldots, 1)$.

Now we construct a continuous function h on the interval I satisfying the following interpolation conditions. At points $\mathcal{X}(q; j_1, \ldots, j_n)$ (the number of such points in $I \backslash E$ is finite)

$$h(\mathcal{X}(q; j_1, \ldots, j_n)) = 2\varepsilon_0 M(q; j_1, \ldots, j_n), \tag{1.11}$$

and

$$h(x) = 2\varepsilon H(x), \qquad x \in E. \tag{1.12}$$

According to the assumptions of the lemma such a function can be chosen from the subspace $Y \subset C(I)$; moreover, in view of the choice of $M(q; j_1, \ldots, j_n)$,

$$\|h\| \leq 2\varepsilon_0 c \max(\|F\|, \|H\|). \tag{1.13}$$

Let $x = (x_1, \ldots, x_n) \in I^n$. If $x \in P_q(j_1, \ldots, j_n)$, then

$$\begin{aligned} h(\psi_q(x_1, \ldots, x_n)) =& h(\mathcal{X}(q; j_1, \ldots, j_n)) = 2\varepsilon_0 M(q; j_1, \ldots, j_n) \\ =& 2\varepsilon F(x_1, \ldots, x_n) + 2\varepsilon_0 [M(q; j_1, \ldots, j_n) F(x_1, \ldots, x_n)] \\ =& 2\varepsilon_0 F(x_1, \ldots, x_n) + \rho_q. \end{aligned}$$

The oscillation of $F(x_1, \ldots, x_n)$ on $P_q(j_1, \ldots, j_n)$ is less than $\varepsilon_0\|F\|$, and hence

$$|\rho_q| \leq 2\varepsilon_0^2 \|F\|.$$

($M(q; j_1, \dots, j_n)$ is the value of F on $P_q(j_1, \dots, j_n)$!) Since every point x in I^n falls into at least $(n+1)$ cubes of the system P_q, $q = 1, \dots, 2n+1$, and $\varepsilon_0 < \dfrac{1}{2(n+1)}$, we shall have

$$\begin{aligned}
&\left| F(x_1, \dots, x_n) - \sum_1^{2n+1} h(\psi_q(x_1, \dots, x_n)) \right| \\
&= \left| F(x_1, \dots, x_n) - \sum_{q=1}^{2n+1} h\left(\sum_{p=1}^{n} \lambda_p \varphi_q(x_p) \right) \right| \\
&\le \left| F(x_1, \dots, x_n) - \sum{}' h(\psi_q(x_1, \dots, x_n)) \right| + \left| \sum{}'' h(\psi_q(x_1, \dots, x_n)) \right|,
\end{aligned}$$

where $\sum'$ is taken over $(n+1)$ of the q's for which cubes from the series I_q cover the point $(x_1, \dots, x_n)$, and $\sum''$ contains the n remaining q's. Let us continue the above estimate:

$$\begin{aligned}
&\le |F(x_1, \dots, x_n) - 2\varepsilon_0(n+1) F(x_1, \dots, x_n)| + 2\varepsilon_0^2(n+1)\|F\| + n\|h\| \\
&< (1 - 2\varepsilon_0(n+1))\|F\| + 2\varepsilon_0^2(n+1)\|F\| + 2n\varepsilon_0 c \max(\|F\|, \|H\|) \\
&\le \left(1 - 2\varepsilon_0(n+1) + 2\varepsilon_0^2(n+1) + 2n\varepsilon_0\right) c \max(\|F\|, \|H\|) \\
&\le (1 - \varepsilon_0) c \max(\|F\|, \|H\|).
\end{aligned}$$

Thus, the set of functions $(\varphi_1, \dots, \varphi_{2n+1}) \in \Phi^{2n+1}$ we have constructed belongs to $\Omega(F, H)$ and at the same time to $\mathcal{P}$. We have proved that $\Omega(F, H)$ is an everywhere dense set in Φ^{2n+1}.

DISCUSSION. Clearly, the crux of the above construction was the possibility of guaranteeing that for an arbitrary point from I^n the number of series of cubes covering it is larger than the number of those that miss it. According to a well-known theorem of Lebesgue from basic dimension theory (cf. [**72a**] for the principles of the theory), if "small" cubes cover I^n, then there is a point in I^n that is covered by $(n+1)$ cubes. So, less than $(n+1)$ series cannot possibly be enough to cover I^n, although we needed $2n+1$ such series.

Another important point was the rational independence of $\lambda_1, \dots, \lambda_n$ that provided injectivity of the inner product $\langle \lambda, a \rangle : a \in I^n \to \langle \lambda, a \rangle \in I$ on a set of vectors with rational coordinates. It is plausible that when choosing $\lambda_1, \dots, \lambda_n$ one could require only injectivity of $\langle \lambda, a \rangle$ on vectors a whose coordinates consist exclusively of zeros or ones, and then construct an everywhere dense set of real numbers containing zero and one such that the inner product $\langle \lambda, a \rangle$ is injective on vectors a with coordinates from that set.

Finally, let us draw attention to the importance of constructing the values of $\psi_q(x_1, \dots, x_n)$ in such a way that those values did not fall into the set E. This required a whole new set of tricks that would not have been necessary for the proof of Theorem 1.1.

COMPLETION OF THE PROOF OF THEOREM 1.2. Now let $A = \{F_\mu\}$, $\mu = 1, 2, \dots$, be a countable, everywhere dense family of functions in $C(I^n)$, and let $B = \{H_\nu\}$, $\nu = 1, 2, \dots$, be a countable everywhere dense family in $C(E)$. We can assume that whenever H_ν, H_{ν_1} belong to B, then so do $H_\nu \pm H_{\nu_1}$ (e.g., B could be the set of all polynomials with rational coefficients). Set $\Omega = \bigcap_{\mu,\nu} \Omega(F_\mu, H_\nu)$. Clearly, Ω contains quasi-all collections $(\varphi_1, \dots, \varphi_{2n+1})$ in Φ^{2n+1}. For each $f(x_1, \dots, x_n) \in C(I^n)$, each arbitrary set $(\varphi_1, \dots, \varphi_{2n+1}) \in \Omega$, and every $H \in C(E)$, let us show that there exists a function $g \in Y$ such that (1.1) holds and

$$g(x) = 2\varepsilon_0 H(x). \tag{1.14}$$

Obviously, (1.3) and (1.14) are the same condition. Choose a number ε_1 so that $0 < \varepsilon_1 < \varepsilon_0$, but $(1-\varepsilon_1)c < 1$. For every $f \in C(I^n)$, it is possible to find a function $F \in A$ such that

$$\|f - F\| < (\varepsilon_0 - \varepsilon_1)\|f\|, \qquad \|F\| \le \|f\|. \tag{1.15}$$

Taking an arbitrary function $H \in B$, and applying Lemma 1.6 to F and H, we find $h(x) \in Y$ satisfying the following conditions:

$$\begin{gathered} \left\| f(x_1, \dots, x_n) - \sum_{q=1}^{2n+1} h\left(\sum_{p=1}^{n} \lambda_p \varphi_q(x_p)\right) \right\| < (1-\varepsilon_1)c\max(\|f\|, \|H\|), \\ h(x) = 2\varepsilon_0 H(x), \quad x \in E, \qquad \|h\| \le 2\varepsilon_0 c \max(\|f\|, \|H\|). \end{gathered} \tag{1.16}$$

We shall denote such a function by $h = \gamma(f, H)$ (there are many such h's for a given f and H, so we take any one of them). Finally, let us take an arbitrary function $H(x) \in C(E)$ and expand it into the series

$$\begin{gathered} H(x) = H_0(x) + H_1(x) + \cdots + H_m(x) + \cdots, \qquad H_j \in B, \\ \|H_j\| \le [c(1-\varepsilon_1)]^j \beta, \qquad \beta = \max(\|f\|, \|H\|). \end{gathered} \tag{1.17}$$

This can be done by choosing a sequence of functions $\{S_m(x)\} \subset B$ approximating $H(x)$ sufficiently fast and setting $H_m(x) = S_m(x) - S_{m-1}(x)$, $m = 1, 2, \dots$, $H_0(x) = S_0(x)$. Construct recursively the sequences $\{f_j(x_1, \dots, x_n)\}$ and $h_j(x) \in Y$:

$$f_0(x_1, \dots, x_n) = f(x_1, \dots, x_n); \qquad h_j(x) = \gamma(f_j, H_j), \quad j = 0, 1, \dots \tag{1.18}$$

$$f_{j+1}(x_1, \dots, x_n) = f_j(x_1, \dots, x_n) - \sum_{q=1}^{2n+1} h_j\left(\sum_{1}^{n} \lambda_p \varphi_q(x_p)\right). \tag{1.19}$$

In view of (1.16)–(1.17), we obtain the estimates:

$$\begin{aligned} \|f_1\| &\le c(1-\varepsilon_1)\beta; \\ \|f_2\| &\le c(1-\varepsilon_1)\max(\|f_1\|, \|H_1\|) \\ &\le \max\left(c^2(1-\varepsilon_1)^2\beta, c(1-\varepsilon_1)\|H_1\|\right) \le c^2(1-\varepsilon_1)^2\beta. \end{aligned}$$

In general,

$$\|f_j\| \le [c(1-\varepsilon_1)]^j \beta, \qquad j = 1, 2, \dots. \tag{1.20}$$

The norms of the h_j can be estimated similarly:

$$\begin{aligned} \|h_0\| &\le 2\varepsilon_0 c \max(\|f\|, \|H_0\|) = 2\varepsilon_0 c\beta, \\ \|h_1\| &\le 2\varepsilon_0 c \max(\|f_1\|, \|H_1\|) \le 2\varepsilon_0 c\beta(c(1-\varepsilon_1)), \\ \|h_2\| &\le 2\varepsilon_0 c \max(\|f_2\|, \|H_2\|) \le 2\varepsilon_0 c\beta[c^2(1-\varepsilon_1)^2]. \end{aligned}$$

In general,

$$\|h_j\| \leq 2\varepsilon_0 c\left[c\left(1-\varepsilon_1\right)\right]^j, \qquad j=1,2,\ldots. \tag{1.21}$$

In view of the estimates (1.20) and (1.21), the series

$$h_0+h_1+\cdots+h_m+\cdots \tag{1.22}$$

converges in norm in $C(I)$, while $f_m \to 0$ in $C\left(I^n\right)$ when $m \to \infty$. Let $g(x)$ and $g_m(x)$ denote the n-th partial sum and the sum of the series (1.22), respectively. Adding up the equalities (1.19) for $j=0,\ldots,m$, we obtain

$$f_{m+1}\left(x_1,\ldots,x_n\right)=f_0\left(x_1,\ldots,x_n\right)-\sum_{q=1}^{2n+1} g_m\left[\sum_{p=1}^{n} \lambda_p \varphi_q\left(x_p\right)\right].$$

Taking the limit in the last equality as $m \to \infty$, we obtain (1.1). Moreover, $g \in Y$ since $g_m \in Y$, $m=1,2,\ldots$, and Y is a closed subspace in $C(I)$. Also, the interpolating property (1.12) holds according to the construction of the functions h_j. The proof of the theorem is now complete.

REMARK. Instead of quasi-all collections $\left(\varphi_1,\ldots,\varphi_{2n+1}\right) \in \Phi^{2n+1}$ in Theorem 1.1, one could consider quasi-all collections from $[C(I)]^{2n+1}$. The proof remains the same.

A geometric interpretation of Kolmogorov's theorem. Let $\lambda_1,\ldots,\lambda_n$ be the same as in Theorems 1.1 and 1.2, and let $\left(\varphi_1,\ldots,\varphi_{2n+1}\right)$ be a collection of functions from Φ^{2n+1} that provides the possibility of representation (1.1). Define a continuous embedding of I^n into $\mathbb{R}^{2n+1}$ by setting

$$X_q=\sum_{p=1}^{n} \lambda_p \varphi_q\left(x_p\right), \qquad p=1,\ldots,2n+1. \tag{1.23}$$

The image of I^n in $\mathbb{R}^{2n+1}$ under this map is a compact set Γ inside I^{2n+1}. According to (1.1), for an arbitrary $f \in C\left(I^n\right)$ we have the following representation:

$$f\left(x_1,\ldots,x_n\right)=\sum_{1}^{2n+1} g\left(X_q\right). \tag{1.24}$$

This, in particular, implies that the embedding (1.23) is a homeomorphism. Indeed, if two different points $x^1=\left(x_1^1,\ldots,x_n^1\right)$ and $x^2=\left(x_1^2,\ldots,x_n^2\right)$ correspond to the same point $\left(X_1,\ldots,X_{2n+1}\right)$, then, taking a function f so that it assumes different values at x^1 and x^2, we would have been unable to represent it by (1.24). We can interpret (1.1) and (1.24) as formulas that allow us to extend to I^{2n+1} an arbitrary (continuous on Γ) function by the formula

$$\sum_{1}^{2n+1} g\left(X_q\right), \qquad g \in C(I). \tag{1.25}$$

In other words, the subspace $Z \subset \left[\left(I^{2n+1}\right)\right]$ that consists of all functions (1.25) interpolates freely on Γ. Thus, Kolmogorov's Theorem admits the following geometric interpretation: there exists a homeomorphic embedding of I^n into I^{2n+1} such that the subspaces Z of type (1.25) freely interpolate a compact set Γ, the image of I^n.

Ostrand [**114**] and Tikhomirov [**87**] extended Kolmogorov's Theorem to arbitrary n-dimensional metric compact sets. Namely, if K is an n-dimensional compact set, then there exists a homeomorphic embedding $\varphi : K \to I^{2n+1}$ ($X_q = \varphi_q(x)$, $x \in K$, $q = 1, \ldots, 2n+1$, $\varphi_q(x) \in C(K)$, $\varphi(x) = (\varphi_1(x), \ldots, \varphi_{2n+1}(x))$) such that the subspace of functions (1.25) interpolates freely on the compact $\Gamma = \varphi(K)$ (cf. Theorem 2.15 below).

The fact that any n-dimensional compact set K can be homeomorphically embedded into I^{2n+1} had been known much earlier. This is the Menger–Nöbeling theorem (cf. [**72a**], p.84). However, the Kolmogorov-Ostrand-Tikhomirov theorem gives much more, since the compact set Γ is shown to possess an important additional property.

Let us make a remark concerning the geometric structure of Γ. Consider a continuous curve γ in $\mathbb{R}^{2n+1}$ defined by the parametric equations

$$X_q = \varphi_q(t), \qquad 0 \le t \le 1, \quad q = 1, \ldots, 2n+1. \tag{1.26}$$

Formula (1.23) for coordinates of points of the compact Γ shows that Γ is a convex combination with coefficients $\lambda_1, \ldots, \lambda_n$ of n copies of the curve γ.

Fridman's improvement. Quasi-all collections $(\varphi_1, \ldots, \varphi_{2n+1}) \in \Phi^{2n+1}$ consist of strictly increasing functions. So, we can assume that functions $\varphi_1, \ldots, \varphi_{2n+1}$ in Theorem 1.1 are strictly increasing. Consider a curve γ given by (1.26) with such φ_q, so it is in this case a simple arc. Since the functions φ_q have bounded variation (they are increasing), γ is a rectifiable curve. Let s be the arc length parameter on γ showing the length of the arc corresponding to the segment $[0, t]$ of the parameter t, and let S denote the total length of γ. Setting $\sigma = \dfrac{s}{S}$, we can define γ by the natural parametric equations

$$X_q = \varphi_q(\sigma), \qquad 0 \le \sigma \le 1, \quad q = 1, \ldots, 2n+1. \tag{1.27}$$

Let $\sigma_1 = \dfrac{s_1}{S_1}$, $\sigma_2 = \dfrac{s_2}{S_2}$ be two values of the parameter, and X_q^1, X_q^2 the corresponding values of the coordinate function X_q. Clearly,

$$\left|X_q^1 - X_q^2\right| = \left|\varphi_q(\sigma_1) - \varphi_q(\sigma_2)\right| \le |s_1 - s_2| \le S\,|\sigma_1 - \sigma_2|, \tag{1.28}$$

so the functions φ_q satisfy the Lipschitz condition of order one.

Since $\Gamma = \sum_1^n \lambda_p \gamma_p$, where γ_p are copies of γ, we have for $(X_1, \ldots, X_{2n+1}) \in \Gamma$

$$X_q = \sum_{p=1}^{n} \lambda_p \varphi_q(\sigma_p). \tag{1.29}$$

Therefore, we have the maps

$$\begin{array}{ccc} (x_1, \ldots, x_n) & \longleftrightarrow & (\sigma_1, \ldots, \sigma_n) \\ & \nwarrow\!\searrow \quad \swarrow\!\nearrow & \\ & (X_1, \ldots, X_{2n+1}) & \end{array}$$

where the diagonal arrows are homeomorphisms of I^n onto Γ, while the horizontal one is a homeomorphism of I^n onto itself. The functions $f(x_1, \ldots, x_n) \in C(I^n)$

pass into the function $F(\sigma_1, \dots, \sigma_n) \in C(I^n)$, and for any $F(\sigma_1, \dots, \sigma_n) \in C(I^n)$ we have

$$F(\sigma_1, \dots, \sigma_n) = \sum_{q=1}^{2n+1} g\left(\sum_{p=1}^{n} \lambda_p \varphi_q(\sigma_p)\right). \tag{1.30}$$

Thus, for any $F \in C(I^n)$, the functions φ_q in the representation (1.30) can be chosen to be strictly increasing and to satisfy the Lipschitz condition of order one.

The possibility of choosing Lipschitz functions for inner functions in Kolmogorov superpositions was first established by Fridman [**52**]. It required a significant improvement of Kolmogorov's original construction. An observation showing that in fact such a possibility follows automatically from Kolmogorov's original statement is due to Kahane [**74**]. Some deep investigations of Vitushkin [**143–146**] show that if the φ_q are continuously differentiable, then the representation (1.30) is impossible for certain $f \in C(I^n)$.

§2. Duality in problems concerning representations by superpositions

From now on $C(T)$ denotes the Banach space of real-valued continuous functions on a compact Hausdorff space T. The space $C(T)$ is equipped with the usual uniform norm (if $g \in C(T)$, then $\|g\| = \max|g(t)|, t \in T$). By $B(T)$ we denote the Banach space of real-valued bounded functions on an arbitrary set T. In that case, for $g(t) \in B(T)$, $\|g\| = \sup|g(t)|$, $t \in T$.

In connection with the contents of Section 1, it is natural to consider the following problem. Let $X, X_1, \dots, X_N$ be compact sets, and $\varphi_i : X \to X_i$ continuous maps. In $C(X)$ let us consider the subspace D consisting of the following functions:

$$D = \{g_1 \circ \varphi_1(x) + \dots + g_N \circ \varphi_N(x)\}, \qquad g_i \in C(X_i), \quad i = 1, \dots, N. \tag{2.1}$$

What conditions should be imposed on X, $\{X_i\}$, $\{\varphi_i\}$ in order to guarantee $D = C(X)$? In other words, under what conditions can one claim that for each $f(x) \in C(X)$ there is a representation

$$f(x) = g_1 \circ \varphi_1(x) + \dots + g_N \circ \varphi_N(x), \qquad g_i \in C(X_i)? \tag{2.2}$$

The same question appears when one considers more general subspaces D. Let the functions $h^i(x) \in C(X)$, $h^i(x) \not\equiv 0$, be given, and consider the subspace D given by

$$D = \left\{h^1(x) g_1 \circ \varphi_1(x) + \dots + h^N(x) g_N \circ \varphi_N(x), \quad g_i \in C(X_i) i = 1, \dots, N\right\}. \tag{2.3}$$

Again, we raise the question of what conditions would imply $D = C(X)$. Instead of finite sums in the definition of D, one could also use infinite series.

Let the compact sets X, $\{X_i\}$, $i = 1, \dots,$ continuous maps $\varphi_i : X \to X_i$, and the functions $h^i(x) \in C(X)$, $h^i(x) \not\equiv 0$, be given. Consider the subspace $D \subset C(X)$ that consists of functions

$$f(x) = \sum_{i=1}^{\infty} h^i(x) g_i(x) \circ \varphi_i(x), \tag{2.4}$$

where $g_i \in C(X_i)$ and the series

$$\sum_i \|h^i\| \, \|g_i\| < +\infty \tag{2.5}$$

converges. (In the case when the sequence $\{X_i\}$ is finite, (2.5) does not impose any additional restrictions on the selection of $g_i \in C(X_i)$.) Functions (2.3) and (2.4) are called linear superpositions. The given functions $h^i(x)$ are called basis functions, while the functions g_i are called coefficients. Thus, the question that interests us can be formulated in the following way: Under what hypotheses does a subspace of linear superpositions coincide with $C(X)$?

In the case when all basis functions $h^i(x) \equiv 1$, the latter question, in view of the well-known Stone theorem (see, e.g., [**40**]), can be interpreted in terms of function algebras: in $C(X)$ there are given closed subalgebras $A_1, \dots, A_N$, each one of which contains the constants. Under what conditions does their algebraic sum

$$A_1 + \cdots + A_N \tag{2.6}$$

coincide with $C(X)$ (cf. §1 in Chapter 2 for more details)? In this form the question readily extends to the case of a countable set of such algebras.

Subspaces similar to D can be constructed out of merely bounded functions as well. Let X, $\{X_i\}$ be arbitrary sets, $\varphi_i : X \to X_i$ arbitrary maps, $h^i(x) \in B(X_i)$, $h^i(x) \not\equiv 0$, given functions, $i = 1, \dots N$, or $i = 1, 2, \dots$. In the space $B(X)$ consider the subspace BD defined by one of the following:

$$BD = \{g_1 \circ \varphi_1(x) + \cdots + g_N \circ \varphi_N(x)\}, \qquad g_i \in B(X_i), \quad i = 1, \dots, N; \tag{2.7}$$

$$\begin{gathered} BD = \left\{h^1(x) g_1 \circ \varphi_1(x) + \cdots + h^N(x) g_N \circ \varphi_N(x)\right\}, \\ g_i \in B(X_i), \quad i = 1, \dots, N; \end{gathered} \tag{2.8}$$

$$BD = \left\{ \sum_{i=1}^{\infty} h^i(x) g_i(x) \circ \varphi_i(x) \right\}, \qquad g_i \in B(X_i), \quad i = 1, 2, \dots, \tag{2.9}$$

where the g_i satisfy (2.5). The question that interests us then is, under what conditions do we have $BD = B(X)$?

The above questions turn out to be dual to the problem of distortion of certain classes of measures under the mappings φ_i. On X, let there be given a real, finitely additive measure μ defined on an algebra $\mathcal{M}$ of subsets of X and having total variation $\|\mu\|$. (In the sequel we only consider measures with finite total variation, and for such a measure μ, $\|\mu\|$ denotes its total variation.) Let $\Phi : X \to Y$ be a mapping of the set X into a set Y. Let $\nu = \Phi \circ \mu$ denote the measure defined on the algebra $\mathcal{N}$ of subsets of Y such that $E \in \mathcal{N}$ if $\Phi^{-1}(E) \in \mathcal{M}$, while

$$\nu(E) = \mu\left(\Phi^{-1}(E)\right). \tag{2.10}$$

Clearly,

$$\|\Phi \circ \mu\| \le \|\mu\|, \tag{2.11}$$

since under the mapping Φ there is a possibility of mixing up the images of those sets on which μ is positive with those where it is negative.

It turns out that the answers to the questions raised above concerning coincidence of D with $C(X)$ or BD with $B(X)$ are associated with whether the maps $\{\varphi_i\}$ provide, for an arbitrary measure μ in a certain class, not too large a damage from such mixing for at least one φ_i. Note that if a bounded function $g(y)$ defined on Y is measurable with respect to the algebra $\mathcal{N}$, then $g[\Phi(x)]$ is measurable with respect to $\mathcal{M}$ and

$$\int_X g[\Phi(x)]\,d\mu = \int_Y g(y)d[\Phi\circ\mu]. \tag{2.12}$$

THEOREM 2.1. *In order that the subspace D of functions* (2.4) *satisfying* (2.5) *coincide with $C(X)$, it is necessary and sufficient that there exists a number λ, $0<\lambda\le 1$, such that for an arbitrary, regular Borel measure μ on X,*

$$\sup_i \frac{\|\varphi_i\circ(h^i\mu)\|}{\|h^i\|} \ge \lambda\|\mu\|. \tag{2.13}$$

PROOF. First, recall that the dual space $C(T)^\star$ of the space $C(T)$ (T is compact) can be identified with the space of regular Borel measures on T with the norm equal to the total variation of measures. (This is the well-known theorem of F. Riesz, see, e.g., [**40,** Chapter IV].) Another important needed fact from functional analysis is the following. Let V and W be Banach spaces, $V^\star$ and $W^\star$ their duals, $A: V\to W$ a continuous linear operator, and $A^\star: W^\star\to V^\star$ its adjoint. In order that the operator A be surjective, it is necessary and sufficient that there exists a positive number λ such that $\|A^\star\mu\|\ge\lambda\|\mu\|$ for all $\mu\in W^\star$. In order that the operator $A^\star$ be surjective, it is necessary and sufficient that there exists a number $\lambda>0$ such that $\|Av\|\ge\lambda\|v\|$ for all $v\in V$. (The latter part of this criterion is not needed here, but will be used later on in this section.)

To prove Theorem 2.1, we argue as follows. Let $G_1,\dots,G_i,\dots$ be a sequence of Banach spaces, and let $\{\eta_i\}$ be a sequence of positive numbers. Consider a subspace G, whose elements g are sequences $\{g_i\}$, $g_i\in G_i$, $i=1,\dots,$ such that

$$\|g\| \stackrel{\text{def}}{=} \sum_1^\infty \eta_i\|g_i\| < +\infty. \tag{2.14}$$

With the norm (2.14) G becomes a Banach space. The dual space $G^\star$ of G can be identified with the space of sequences $L=\{L_i\}$ of linear functions $L_i\in G_i^\star$, $i=1,\dots,$ and

$$\|L\| = \sup_i \frac{\|L_i\|}{\eta_i}. \tag{2.15}$$

(Obviously, only those sequences L are included in $G^\star$ for which $\|L\|$ in (2.15) is finite.) A functional L acts on elements $g=\{g_i\}$ as follows:

$$\langle g, L\rangle = L(g) = \sum_i L_i(g_i) = \sum_i\langle g_i, L_i\rangle. \tag{2.16}$$

The series (2.16) converges absolutely in view of (2.14) and (2.15).

In the context of Theorem 2.1 we take $G_i = C(X_i)$, $i=1,\dots,$ $\eta_i = \|h^i\|$, and construct the space G as above. Define a continuous linear operator $A: G\to C(X)$ by associating to each $g\in G$, $g=(g_1,\dots,g_i,\dots)$, $g_i\in C(X_i)$, a function $Ag = f(x)$ defined by (2.4). In view of our construction and (2.14), the series (2.4) converges in $C(X)$ and $\|A\|\le 1$. Find the adjoint $A^\star: C(X)^\star\to G^\star$. For an arbitrary regular Borel measure μ on X (defining a continuous functional μ on $C(X)$) and any $g \in G$ we must have

$$\langle Ag, \mu \rangle = \langle g, A^\star \mu \rangle.$$

From (2.4), (2.12), and (2.16) it follows that

$$\langle Ag, \mu \rangle = \int_X \left(\sum_i h^i(x) g_i \circ \varphi_i(x) \right) d\mu = \sum_i \int_{X_i} g_i d \left[\varphi_i \circ h^i \mu \right].$$

Hence,

$$A^\star \mu = \left(\dots, \varphi_i \circ \left(h^i \mu \right), \dots \right). \tag{2.17}$$

Since by (2.15) with $\eta_i = \left\| h^i \right\|$ we have

$$\| A^\star \mu \| = \sup_i \frac{\left\| \varphi_i \circ \left(h^i \mu \right) \right\|}{\| h^i \|}, \tag{2.18}$$

the criterion for surjectivity of the operator A has the form (2.13). Since

$$\left\| \varphi_i \circ \left(h^i \mu \right) \right\| \leq \left\| h^i \mu \right\| \leq \left\| h^i \right\| \| \mu \|,$$

we obtain for λ in (2.13) the estimate $0 < \lambda \leq 1$. Theorem 2.1 is proved.

The space $B(T)^\star$, dual to $B(T)$ (T is an arbitrary set), consists of finitely additive measures of bounded variation defined on the algebra of all subsets ([**40,** Chapter IV]). So, introducing some obvious changes in the above arguments we obtain the following result.

THEOREM 2.2. *In order that the subspace BD of functions* (2.9) *satisfying* (2.5) *coincide with $B(X)$, it is necessary and sufficient that there exists a number λ, $0 < \lambda \leq 1$, such that* (2.13) *holds for any measure $\mu \in B(X)^\star$.*

Finitely additive measures that are elements of $B(X)^\star$ do not represent a convenient object. Hence, Theorem 2.2 is not very useful. There is a more convenient result in this direction that deals with a significantly simpler class of measures $\ell^1(T)$, the subspace of discrete measures on T. So, $\mu \in \ell^1(T)$ means that

$$\mu = \sum_1^\infty \beta_i \delta_{t_i}, \qquad \| \mu \| = \sum_1^\infty |\beta_i| < +\infty, \tag{2.19}$$

where $\{t_i\}$ is a (countable) sequence of points in T, $\{\beta_i\}$ is a sequence of real numbers, and δ_t is a delta-mass at point t. It is known that $B(T) = \ell^1(T)^\star$ (cf. [**40,** Chapter IV]).

THEOREM 2.3. *Let the assumptions of Theorem 2.2 regarding X, $\{X_i\}$, $\{\varphi_i\}$ hold, let*

$$d \stackrel{\text{def}}{=} \sum_i \left\| h^i \right\| < +\infty, \tag{2.20}$$

and let BD consist of the functions (2.9) *for which instead of* (2.5) *the condition*

$$\sup_i \| g_i \| < +\infty \tag{2.21}$$

holds. In order that $BD = B(X)$, it is necessary and sufficient that there exists a number λ, $0 < \lambda \leq d$, such that for any $\mu \in \ell^1(X)$ the following inequality holds:

$$\sum_1^\infty \left\|\varphi_i \circ \left(h^i \mu\right)\right\| \geq \lambda \|\mu\|. \tag{2.22}$$

PROOF. First, note that the series (2.22) converges. Indeed,

$$\sum_1^\infty \left\|\varphi_i \circ \left(h^i \mu\right)\right\| \leq \sum_1^\infty \left\|h^i \mu\right\| \leq \|\mu\| \sum_1^\infty \left\|h^i\right\| = d\|\mu\|, \tag{2.23}$$

and we obtain an estimate for the number λ.

Now consider a Banach space E of sequences $\nu = (\nu_1, \dots, \nu_i, \dots)$, where $\nu_i \in \ell^1(X_i)$, $i = 1, \dots,$ such that

$$\|\nu\| \stackrel{\text{def}}{=} \sum_1^\infty \|\nu_i\| < +\infty. \tag{2.24}$$

Then the dual space $E^\star$ consists of all sequences $g = (g_1, \dots, g_i, \dots)$, where $g_i \in B(X_i)$ and

$$\|g\| = \sup_i \|g_i\| < +\infty. \tag{2.25}$$

A functional $g \in E^\star$ acts on an element $\nu \in E$ according to the formula

$$\langle \nu, g\rangle = \sum_1^\infty \int_{X_i} g_i d\nu_i. \tag{2.26}$$

In view of (2.24) and (2.25), the series in (2.26) converges absolutely.

Further, consider a continuous linear operator $U : \ell^1(X) \to E$:

$$\mu \in \ell^1(X) \to U\mu = \nu = \left(\nu_1 = \varphi_1 \circ \left(h^1 \mu\right), \dots, \nu_i = \varphi_i \circ \left(h^i \mu\right), \dots\right). \tag{2.27}$$

It indeed follows from the estimate (2.23) that U maps $\ell^1(X)$ into E (and, moreover, $\|U\| \leq d$). Let us calculate the adjoint $U^\star : E^\star \to \ell^1(X)^\star$. For an arbitrary $\mu \in \ell^1(X)$ and an arbitrary $g \in E^\star$, we must have

$$\begin{aligned}\langle \mu, U^\star, g\rangle =& \langle U\mu, g\rangle = \sum_1^\infty \int_{X_1} g_i d\left[\varphi_1 \circ \left(h^i \mu\right)\right] \\ =& \sum_1^\infty \int_X g_i \circ \varphi_i(x) \cdot h^i(x) d\mu = \int_X \left(\sum_1^\infty g_i \circ \varphi_i(x) \cdot h^i(x)\right) d\mu.\end{aligned}$$

So,

$$U^\star g = \sum_1^\infty h^i(x)\left[g_i \circ \varphi_i(x)\right] \in BD. \tag{2.28}$$

In view of (2.24) and the definition of the norm in $E^\star$ in (2.25), the series (2.28) converges absolutely and converges in norm in $B(X)$, and its sum indeed belongs to BD. Now the questions we are interested in, namely when $BD = B(X)$ and when $U^\star(E^\star) = B(X)$, are identical. In other words, they both reduce to the question:

When is the operator $U^\star$ surjective? According to the general functional-analytic criterion cited in the proof of Theorem 2.1, this holds if and only if

$$\|U\mu\| \geq \lambda \|\mu\|, \qquad \forall \mu \in \ell^1(X) \tag{2.29}$$

for some $\lambda > 0$. But in view of (2.27) and (2.24), (2.29) coincides with (2.22).

Consider one more space. Let $c_0(T)$ denote the class of bounded real-valued functions $x(t)$ defined on an (arbitrary) set T such that for any $\varepsilon > 0$ the set $\{t \in T : |x(t)| > \varepsilon\}$ is finite. $c_0(T)$ is a closed subspace in $B(T)$, and $c_0(T)^\star = \ell^1(T)$ ([**35**, Chapter II, Section 2]). We say that a map $\varphi : X \to Y$ has finite rank if the full preimage of any point $y \in Y$ is a finite set. The system of mappings $\varphi_i : X \to X_i$, $i = 1, \dots,$ has finite rank if each mapping φ_i has finite rank.

Let the assumptions of Theorem 2.2 hold in relation to X, $\{X_i\}$, $\{\varphi_i\}$, $\{h^i\}$, and let the system of mappings $\{\varphi_i\}$ have finite rank. Form a subspace c_0D inside $B(X)$ that consists of functions

$$\sum_1^\infty h^i(x) g_i \circ \varphi_i(x), \qquad g_i \in c_0(X_i), \quad i = 1, \dots, \tag{2.30}$$

such that (2.5) holds. According to those assumptions, $c_0D \subset c_0(X)$.

THEOREM 2.4. *Under the above assumptions, $c_0D = c_0(X)$ if and only if one can find a number $\lambda > 0$ such that* (2.13) *holds for all $\mu \in \ell^1(X)$.*

If $h^i(x) \in c_0(X)$, $i = 1, \dots,$ then in the definition of c_0D take an arbitrary $g_i \in B(X_i)$ so that (2.5) holds and the statement of Theorem 2.4 is true even without the assumption concerning the finiteness of the rank of the system $\{\varphi_i\}$.

For the proof one has to repeat the arguments in Theorem 2.1, letting in this case $G_i = c_0(X_i)$. (In Theorem 2.1 we took $G_i = C(X_i)$, while in Theorem 2.2 we had to take $G_i = B(X_i)$). In the case when the subspaces D or BD consist of finite sums, Theorems 2.1–2.4 can be stated in a more unified fashion. It is convenient to separate those results.

THEOREM 2.5. *Let X, X_i, $i = 1, \dots, N$, be compact sets, $\varphi_i : X \to X_i$ continuous mappings, $h^i(x) \in C(X_i)$. For the subspace D of functions* (2.3) *to coincide with $C(X)$ it is necessary and sufficient that there exists a number $\lambda > 0$ such that for all $\mu \in C(X)^\star$ the following holds*:

$$\sup_i \left\|\varphi_i \circ \left(h^i\mu\right)\right\| \geq \lambda \|\mu\|. \tag{2.31}$$

If a subspace D has form (2.1) *(i.e., all $h^i(x) \equiv 1$), then for the equality $D = C(X)$ it is necessary and sufficient that there exists λ, $0 < \lambda \leq 1$, such that*

$$\sup_i \|\varphi_i \circ \mu\| \geq \lambda \|\mu\|. \tag{2.32}$$

THEOREM 2.6. *Let X, X_i, $i = 1, \dots, N$, be arbitrary sets, $\varphi_i : X \to X_i$ arbitrary maps, $h^i(x) \in B(X_i)$, $h^i(x) \not\equiv 0$. For the space* (2.8) *the following statements are equivalent*:

1. *$BD = B(X)$.*
2. *There exists $\lambda > 0$ such that* (2.31) *holds for all $\mu \in B(X)^\star$.*
3. *There exists $\lambda > 0$ such that* (2.31) *holds for all $\mu \in \ell^1(X)$.*

In the case when the subspace BD has the form (2.7), *condition* (2.31) *becomes* (2.32) *with* $0 < \lambda \leq 1$.

Comparing Theorems 2.5 and 2.6, we obtain an interesting corollary.

COROLLARY 2.7. *Let* X, X_i, $i = 1, \dots, N$, *be compact sets, let* $\varphi_i : X \to X_i$ *be continuous mappings, let* $h^i(x) \in C(X_i)$, *and let the subspaces* D *and* BD *be defined by* (2.3) *and* (2.8), *respectively. If* $D = C(X)$, *then* $BD = B(X)$.

PROOF. According to Theorem 2.5, $D = C(X)$, provided that (2.31) holds for all $\mu \in C(X)^\star$. But this implies, of course, that the inequality (2.31) holds for all $\mu \in \ell^1(X)$. Then, Theorem 2.6 states that $BD = B(X)$.

In particular, we obtain free-of-charge the following analogue of Kolmogorov's theorem (Theorem 1.1) for bounded functions.

COROLLARY 2.8. *Let* $\lambda_1 > 0, \dots, \lambda_n > 0$ *be rationally independent numbers,* $\sum_1^n \lambda_i = 1$. *For quasi-all collections* $(\varphi_1, \dots, \varphi_{2n+1}) \in \Phi^{2n+1}$, *the following statement holds: an arbitrary function* $f(x_1, \dots, x_n)$, *bounded on* I^n, *can be represented in the following form:*

$$f(x_1, \dots, x_n) = \sum_{i=1}^{2n+1} g_i \left(\sum_{j=1}^{n} \lambda_j \varphi_i(x_j) \right), \qquad g_i \in B(I). \tag{2.33}$$

Yet, we are unable to guarantee that all the g_i in (2.33) can be replaced by one and the same function g similarly to (1.1). However, the possibility of doing this will be established later (cf. §6 of this chapter).

Naturally, one poses a question converse to Corollary 2.7. Namely, does coincidence of BD and $B(X)$ (under the assumptions of Corollary 2.7) imply coincidence of D and $C(X)$? The answer turns out to be negative. An example will be presented in §8. One may think that if (2.31) holds for all $\mu \in B(X)^\star$, which is necessary for the equality $BD = B(X)$, all the more so should (2.31) follow for all $\mu \in C(X)^\star$, which is equivalent to $D = C(X)$. However, the heart of the matter is that variations for measures in $B(X)^\star$ and those in $C(X)^\star$ are defined differently: in the former case, for arbitrary partitions of X; in the latter, for partitions of X into Borel subsets. Nevertheless, in the direction opposite to Corollary 2.7, one can establish the following result.

THEOREM 2.9. *Under the assumptions of Theorem* 2.1, *let each* $f(x) \in C(X)$ *admit the representation*

$$f(x) = \sum_1^\infty h^i(x) g_i \circ \varphi_i(x), \tag{2.34}$$

where the g_i *are bounded Borel-measurable functions on* X_i, *and* (2.5) *holds. Then* $D = C(X)$, *i.e, for all* $f(x) \in C(X)$ (2.34) *holds with continuous* g_i $(g_i \in C(X_i))$.

PROOF. First of all, note that the set $b(T)$ of bounded Borel-measurable functions on a compact space T is a closed subspace in $B(T)$. Let G be the subspace constructed in the proof of Theorem 2.2. Construct a subspace b in G by defining

$g = (g_1, \dots, g_i, \dots) \in b$ if and only if $g_i \in b(X_i)$, $i = 1, \dots$. The space b is a Banach space. Consider an operator $A : b \to b(X)$ defined by

$$g \mapsto \sum_1^\infty h^i(x) g_i \circ \varphi_i(x), \qquad g_i \in b(X_i). \tag{2.35}$$

By the assumptions, $A(b) \subset C(X)$. Let $\tilde{b} = A^{-1}(C(X))$. Then, $\tilde{b}$ is a closed subspace of b, and so it is a Banach space. Consider the restriction $\tilde{A}$ of A onto $\tilde{b}$: $\tilde{A} : \tilde{b} \to C(X)$. The adjoint operator $\tilde{A}^\star : C(X)^\star \to \tilde{b}^\star$ is defined by formulae (2.17) that were used to define the operator $A^\star$ in Theorem 2.1, since the calculations we made to establish (2.17) also hold for bounded Borel functions g_i, not merely for continuous functions. In addition,

$$L_i(g_i) = \int_{X_i} g_i d\nu_i, \qquad \nu_i = \varphi_i \circ \left(h^i \mu\right) \tag{2.36}$$

is a continuous linear functional on $b(X_i)$ having there the same norm as on the entire $C(X_i)$, i.e., the total variation $\|\nu_i\|$. Let c be a subspace in b generated by the sequences $g = (g_1, \dots, g_i, \dots)$, where $g_i \in C(X_i)$. Clearly, $\tilde{b} \supset c$. If $\tilde{b}(X_i)$ is a natural projection of $\tilde{b}$ onto $b(X_i)$, then it is obvious that $C(X_i) \subset \tilde{b}(X_i) \subset b(X_i)$. Therefore, the norm of the functional (2.36) remains the same on $C(X_i)$ as on $b(X_i)$ and equals the total variation $\|\nu_i\|$. Thus, for $\nu = \tilde{A}^\star \mu = \left(\nu_1, \dots, \nu_i = \varphi_i \circ \left(h^i \mu\right), \dots\right)$ we have $\|\nu\| = \sup_i \dfrac{\|\nu_i\|}{\|h^i\|}$. The equality $\tilde{A}\left(\tilde{b}\right) = C(X)$ holds provided that there exists a number $\lambda > 0$ such that $\left\|\tilde{A}^\star \mu\right\| = \|\nu\| = \sup_i \dfrac{\|\nu_i\|}{\|h^i\|} \geq \lambda \|\mu\|$. So, for all $\mu \in C(X)^\star$ condition (2.13) holds, and hence, according to Theorem 2.1, $D = C(X)$.

The contents of this section are based on the work of Sternfeld [**133, 134**]. (1),(3) of Theorem 2.6 when $h^i = 1$, $i = 1, \dots, N$, and Corollary 2.7 are proved in [**133**]. Theorem 2.5 (also, for $h^i \equiv 1$) is proved in [**134**]. The more general formulations presented here did not require any new ideas.

Theorem 2.9 with finitely many terms and $h^i \equiv 1$ is Theorem 4 from [**131**]. However, its proof there is based on the erroneous Theorem 3 in that paper. A counterexample to Theorem 3 from [**131**] is due to V. A. Medvedev (cf. §8).

§3. Separation of points and measures

Various types of point separation. We have already noted in the previous section that the criteria obtained there are associated with a certain separation under mappings φ_i of images of sets on which a measure (from some class $C(X)^\star$, $B(X)^\star$, $\ell^1(X)$) is positive from those where it is negative. In this section we are going to study this phenomenon in detail.

Let X be a set and $F = \{\varphi\}$ be a family of functions defined on X. In general, each one of these functions may have its own range, and the precise nature of those ranges is unimportant for now. The family $F = \{\varphi\}$ separates points in X if for any two points $x_1 \in X$, $x_2 \in X$, $x_1 \neq x_2$, there exists φ in F such that $\varphi(x_1) \neq \varphi(x_2)$. If F consists of scalar-valued functions, then we say that F *strictly* separates points in X if F separates points in X; and for each $x \in X$, there exists $\varphi \in F$ such that $\varphi(x) \neq 0$ (the latter condition appears in the theory of function algebras).

Again, let elements of F be arbitrary (not necessarily scalar-valued) functions on X. F *strongly* separates points in X if for any two disjoint finite sets with the same cardinality $A = \{x_1, \dots, x_m\}$ and $B = \{y_1, \dots, y_m\}$ (for all m) there exists φ in F so that $\varphi(A) \neq \varphi(B)$. Obviously, strong separation implies separation: it suffices to take $A = \{x_1\}$, $B = \{x_2\}$.

Uniform separation of points. For our theme, however, an even stronger type of separation will be the most important. We say that a family $F = \{\varphi\}$ *uniformly* separates points in X, if there exists a number λ, $0 < \lambda \leq 1$, such that for any two disjoint sequences $x_1, \dots, x_m$ and $y_1, \dots, y_m$ ($\forall m \in N$) of elements of X there exists $\varphi \in F$ satisfying the following property: if we drop from the sequence $\varphi(x_1), \dots, \varphi(x_m)$, $\varphi(y_1), \dots, \varphi(y_m)$ in $\varphi(X)$ the maximal number of pairs $(\varphi(x_i), \varphi(y_j))$ for which $\varphi(x_i) = \varphi(y_j)$, there will remain at least $\lambda 2m$ members of that (joint) sequence of images, and hence the number of pairs removed does not exceed $(1-\lambda)2m$. (Thus, in each of the sequences $\varphi(x_1), \dots, \varphi(x_m)$ and $\varphi(y_1), \dots, \varphi(y_m)$ there will remain at least λm members, i.e., the number of terms removed is at most $(1-\lambda)m$.) Of course, the number of terms in the sequences "thinned out" according to these rules remains the same.

Let us illustrate the removal procedure. Let $\varphi(x_1) = \varphi(x_2) = \alpha$, $\varphi(x_3) = \varphi(x_1) = \beta$, $\beta \neq \alpha$; $\varphi(y_1) = \varphi(y_4) = \alpha$, $\varphi(y_2) = \beta$, $\varphi(y_3) = \gamma$, $\gamma \neq \alpha$, $\gamma \neq \beta$. After removing the coinciding pairs, we have left $\varphi(x_3)$ (or $\varphi(x_4)$) from the first sequence and $\varphi(y_3)$ from the second.

EXAMPLES. 1. Let $X = \{(u,t),\ 0 \leq u,\ t \leq 1\}$ be the unit square and let $F = \{\varphi_1, \varphi_2\}$, where $\varphi_1(u,t) = u$, $\varphi_2(u,t) = t$, be coordinate functions. Clearly, F separates points, but not strongly: taking $A = \{(0,0),(1,1)\}$, $B = \{(0,1),(1,0)\}$, we have $\varphi_1(A) = \{0,1\}$, $\varphi_1(B) = \{0,1\}$, $\varphi_2(A) = \{0,1\}$, $\varphi_2(B) = \{0,1\}$.

2. Let X be the boundary of the triangle in $\mathbb{R}^2$ with vertices $(0,0)$, $(1/2,0)$, and $(1,1)$, while F is the same as in Example 1. Then F strongly separates points in X, but does not separate them uniformly.

Examples for which uniform separation holds will be given later.

It is probably worth pointing out that we had to distinguish between the concepts of a set and a sequence, in order to cover in the case of the latter the possibility of identical elements. At the same time, the order of elements in a sequence (usually quite important) makes no difference to us. In view of this, it is convenient to use the concept of a multi-set.

Multi-sets. A multi-set is a pair (A,a) that consists of a set A and a function $a(x)$ defined on A and taking values in the set of non-negative integers. (Intuitively, $a(x)$ shows how many times the point x appears in A, i.e., the multiplicity of x in A.) The cardinality $|(A,a)|$ of a multi-set (A,a) is defined by

$$|(A,a)| = \sum_{x \in A} a(x). \tag{3.1}$$

The intersection of multi-sets (A,a) and (B,b) is the multi-set $(A \cap B, a \wedge b)$, where

$$(a \wedge b)(x) = \min\{a(x), b(x)\}. \tag{3.2}$$

If (A, a) is a multi-set and φ is a function defined on A, then $\varphi[(A, a)]$ is the multi-set $\varphi[(A, a)] = (\varphi(A), \varphi \circ a)$, where

$$\varphi \circ a(y) = \sum_{x \in \varphi^{-1}(y)} a(x), \quad \text{for } y \in \varphi(A). \tag{3.3}$$

LEMMA 3.1. *A family $F = \{\varphi\}$ uniformly separates points in X if and only if there exists λ, $0 < \lambda \leq 1$, satisfying the following condition. Let (A, a) and (B, b) be two disjoint multi-sets, $|(A, a)| = |(B, b)| = m < \infty$. Then, there exists $\varphi \in F$ satisfying the inequality*

$$|\varphi[(A, a)] \cap \varphi[(B, b)]| \leq (1 - \lambda)m. \tag{3.4}$$

PROOF. Write down the multi-set (A, a) as a sequence $x_1, \dots, x_m$, where each term is repeated as many times as the value of the function $a(x)$. Similarly, represent the set (B, b) by $y_1, \dots, y_m$. Then, the set $\varphi[(A, a)] \cap \varphi[(B, b)] = (\varphi(A) \cap \varphi(B), \ \varphi \circ a \wedge \varphi \circ b)$ is precisely the set that is removed from $\varphi(x_1), \dots, \varphi(x_m)$ and $\varphi(y_1), \dots, \varphi(y_m)$ according to the process described in the definition of uniform separation of points. So, if (3.4) holds, then after the removal, each one of the sequences $\varphi(x_1), \dots, \varphi(x_m)$ and $\varphi(y_1), \dots, \varphi(y_m)$ will contain at least λm terms. Conversely, if there are λm or more terms left, then for the number of terms removed (3.4) must hold.

Separation of measures. Let X be a set, $\mathcal{M}$ an algebra of subsets of X, $(X, \mathcal{M})$ a measure space. For a real-valued (finitely additive) measure μ on $(X, \mathcal{M})$, $\|\mu\|$ will denote the total variation of μ. We shall only consider measures with finite variations. If $\varphi : X \to Y$ and $\mathcal{N} = \{E \subset Y : \varphi^{-1}(E) \in \mathcal{M}\}$, then the measure $\varphi \circ \mu$ on $(Y, \mathcal{N})$ is defined by

$$\varphi \circ \mu(E) = \mu\left(\varphi^{-1}(E)\right), \qquad E \in \mathcal{N}. \tag{3.5}$$

Clearly, $\|\varphi \circ \mu\| \leq \|\mu\|$. Formulas (3.4) and (3.5) show that under a mapping φ multiplicities in multi-sets and measures are transformed according to one and the same rule.

Let S be a class of measures μ defined on X and $F = \{\varphi\}$ be a family of functions defined in X. We shall say that F uniformly separates measures of class S if there exists λ, $0 < \lambda \leq 1$, such that for each μ in S one can find $\varphi \in F$ so that

$$\|\varphi \circ \mu\| \geq \lambda \|\mu\|. \tag{3.6}$$

The fact that (3.6) does define some sort of "separation" is intuitively obvious. Indeed, in forming the values of $\varphi \circ \mu$, the values of μ are "mixed", and to the value of $\varphi \circ \mu$ on a set E there may be contributions from values of μ on disjoint sets $\varphi^{-1}(E)$, on some of which μ is positive, on others, negative. Hence, the decreasing of $\|\varphi \circ \mu\|$ in comparison with $\|\mu\|$. The inequality (3.6), on the other hand, states that such "mixing" is not very significant. More precisely, this is seen from the following. Let A and B be two disjoint sets in $\mathcal{M}$ such that μ takes positive values on subsets of A and negative values on subsets of B: $\mu(A_1) \geq 0$ for all $A_1 \subset A$, $A_1 \in \mathcal{M}$; $\mu(B_1) \leq 0$ for all $B_1 \subset B$, $B_1 \in \mathcal{M}$; $\mu(H) = 0$ for all $H \in \mathcal{M}$, $H \cap (A \cup B) = \emptyset$. Set $C = \varphi(A) \cap \varphi(B)$, $D = \varphi^{-1}(C) \cap A$, $E = \varphi^{-1}(C) \cap B$, $A_1 = A \backslash D$, $A' = \varphi(A_1) = \varphi(A) \backslash C$, $B_1 = B \backslash E$, $B' = \varphi(B_1) = \varphi(B) \backslash C$. The term "separation" will be justified if we show that some characteristic of the set

C in terms of the measure μ is small. Naturally, we assume that all of the sets introduced above belong to $\mathcal{M}$ or $\mathcal{N}$, accordingly. In particular, this imposes certain restrictions on the properties of X, $\mathcal{M}$, φ, and μ. For $P \subset C$, let $P' = \varphi^{-1}(P) \cap D$, $P'' = \varphi^{-1}(P) \cap E$, and

$$\Theta(P) \stackrel{\text{def}}{=} \min\left(\mu\left(P'\right), -\mu\left(P''\right)\right). \tag{3.7}$$

Consider a partition of C into subsets P_i, $i = 1, \dots, m$, such that $P_i \cap P_j = \emptyset$, $i \neq j$, $\bigcup_{j=1}^{m} P_i = C$, and define the number Θ by

$$\Theta = \inf \sum_{1}^{m} \Theta\left(P_i\right), \tag{3.8}$$

where the infimum is taken over all partitions of set C. Clearly, Θ represents a certain characteristic in terms of the measure μ of the quantity $\varphi(A) \cap \varphi(B)$.

LEMMA 3.2. *With φ, μ and Θ as above,*

$$\|\varphi \circ \mu\| = \|\mu\| - 2\Theta. \tag{3.9}$$

PROOF. We have

$$\begin{aligned}\|\varphi \circ \mu\| &= \|\varphi \circ \mu\|_{A'} + \|\varphi \circ \mu\|_{C} + \|\varphi \circ \mu\|_{B'} \\ &= \varphi \circ \mu\left(A'\right) + \|\varphi \circ \mu\|_{C} - \varphi \circ \mu\left(B'\right) = \mu\left(A_1\right) - \mu\left(B_1\right) + \|\varphi \circ \mu\|_{C},\end{aligned}$$

$$\|\varphi \circ \mu\|_{C} = \sup \sum_{1}^{m} \left|\varphi \circ \mu\left(P_i\right)\right| = \sup \sum_{1}^{m} \left|\mu\left(P_i'\right) + \mu\left({P''}_{\mu}\right)\right|$$

(sup is taken over all partitions of the set C, $P_i' = \varphi^{-1}\left(P_i\right) \cap D$, $P_i'' = \varphi^{-1}\left(P_i\right) \cap E$. The P_i' give a partition of D, while the P_i'' give a partition of E). Continuing the above equality, we have

$$\begin{aligned}\|\varphi \circ \mu\|_{C} &= \sup \sum_{1}^{m} \left(\mu\left(P_i'\right) - \mu\left(P_i''\right) - 2\Theta\left(P_i\right)\right) \sup \left[\mu\left(D\right) - \mu(E) - 2\sum_{1}^{m} \Theta\left(P_i\right)\right] \\ &= \mu\left(D\right) - \mu(E) - 2\inf \sum_{1}^{m} \Theta\left(P_i\right) = \mu\left(D\right) - \mu(E) - 2\Theta.\end{aligned}$$

From the latter equality we obtain

$$\begin{aligned}\|\varphi \circ \mu\| &= \mu\left(A_1\right) + \mu\left(D\right) - \left(\mu\left(B_1\right) + \mu(E)\right) - 2\Theta \\ &= \mu(A) - \mu(B) - 2\Theta = \|\mu\| - 2\Theta.\end{aligned}$$

From this we immediately deduce the following analogue of Lemma 3.1.

LEMMA 3.3. *If a family F uniformly separates measures in the class S, $\mu \in S$, and the assumptions of Lemma 3.2 hold, then there exists $\varphi \in F$ such that*

$$2\Theta \leq (1 - \lambda)\|\mu\|. \tag{3.10}$$

If S consists of countably additive measures, then in order for a family of functions F to uniformly separate measures in S, it is necessary and sufficient that there exists λ, $0 < \lambda \leq 1$, such that under the assumptions of Lemma 3.3 the inequality (3.10) holds. Indeed, in the latter case we can use the Hahn decomposition of the measure μ:

$$X = A \cup B, \quad A \cap B = \emptyset, \qquad A_1 \subset A \Rightarrow \mu(A_1) \geq 0, \quad B_1 \subset B \Rightarrow \mu(B_1) \leq 0.$$

In addition to the classes of measures $C(X)^\star$, $B(X)^\star$, $\ell^1(X)$ introduced in §2, we consider several other classes. By $\ell^1_K(X)$ we denote the class of those measures in $\ell^1(X)$ that are supported on finite sets. If S is a class of measures, then by S_0 we denote the subclass of S consisting of those measures $\mu \in S$ for which $\mu(X) = 0$.

LEMMA 3.4. *Let S, S' be two families of measures on a measurable space $(X, \mathcal{M})$, with S' dense in S (with respect to the norm). If a family of functions $F = \{\varphi\}$ uniformly separates measures in S', then it uniformly separates measures in S as well.*

Note that no continuity assumptions are imposed on functions in F.

PROOF. Let $\mu \in S$. For any $\varepsilon > 0$ there exists $\mu' \in S'$ such that $\|\mu - \mu'\| < \epsilon$. Also, there exists $\varphi \in F$ such that $\|\varphi \circ \mu'\| \geq \lambda \|\mu'\|$. We have $\|\varphi \circ \mu - \varphi \circ \mu'\| = \|\varphi \circ (\mu - \mu')\| \leq \|\mu - \mu'\| < \varepsilon$. Hence, $\|\varphi \circ \mu\| \geq \|\varphi \circ \mu'\| - \varepsilon > \lambda \|\mu'\| - \varepsilon > \lambda\|\mu\| - (1 + \lambda)\varepsilon$. Since ε is arbitrary, there exists $\varphi \in F$ satisfying $\|\varphi \circ \mu\| \geq \lambda'\|\mu\|$ for any $\lambda' < \lambda$.

Comparison of uniform separation of points and measures. Let X be an arbitrary set and $F = \{\varphi\}$ be a family of functions on X.

LEMMA 3.5. *The following statements are equivalent.*
1. *F uniformly separates points in X.*
2. *F uniformly separates measures of class $\ell^1_{K,0}$.*
3. *F uniformly separates measures of class $\ell^1_0(X)$.*

PROOF. Let us show that $1 \Rightarrow 2$. Let $\mu \in \ell^1_{K,0}$, and let $x_1, \dots, x_K$ be those points at which atoms μ_i of the measure μ are positive while $y_1, \dots, y_\ell$ are the points where $\mu_i < 0$. In the latter case, set $\nu_i = -\mu_i$. Obviously, in view of Lemma 3.4 it suffices to prove the inequality required in the definition of the separation of measures for the case when all numbers μ_i and ν_i are rational. Find the common denominator D for all those numbers, so $\mu_i = \dfrac{r_i}{D}$, $i = 1, \dots, k$, $\nu_i = \dfrac{R_i}{D}$, $i = 1, \dots, \ell$, where r_i, R_i are positive integers. Since $\mu \in \ell^1_{0,K}(X)$,

$$\sum_1^k r_i - \sum_1^\ell R_i = 0. \tag{3.11}$$

Denote the common value of $\sum_1^k r_i = \sum_1^\ell R_i$ by m and consider the multi-sets (A, a) and (B, b), where $A = \{x_1, \dots, x_k\}$, $a(x_i) = r_i$, $i = 1, \dots, k$; $B = \{y_1, \dots, y_\ell\}$, $b(y_i) = R_i$, $i = 1, \dots, \ell$, so $|(A, a)| = |(B, b)| = m$. Since family F uniformly separates points, there exists a function $\varphi \in F$ satisfying (3.4).

Let us study the measure $\alpha = D\mu$ and its image $\beta = \varphi \circ \alpha$. Atoms of α at points of the set A coincide with the multiplicities r_i of those points, while at points of B they coincide with the multiplicites R_i taken with the minus sign:

$\|\alpha\| = \sum_i^k r_i + \sum_1^\ell R_i = 2m$. Consider a measure β. According to the choice of φ, (3.4) holds. If $z \in \varphi(A)\backslash\varphi(A) \cap \varphi(B)$, then the atom of β at a is given by

$$\beta(z) = \varphi \circ \alpha(z) = \sum_{x_i \in \varphi^{-1}(z)} r_i. \tag{3.12}$$

Similarly, if $z \in \varphi(B)\backslash(A) \cap \varphi(B)$, then

$$\beta(z) = \varphi \circ \alpha(z) = -\sum_{y_i \in \varphi^{-1}(z)} R_i. \tag{3.13}$$

Now, let $z \in \varphi(A) \cap \varphi(B)$ and

$$M \stackrel{\text{def}}{=} \varphi \circ a(z) = \sum_{x_i \in \varphi^{-1}(z)} r_i, \qquad N \stackrel{\text{def}}{=} \varphi \circ b(z) = \sum_{y_i \in \varphi^{-1}(z) \cap B} R_i.$$

Consider the multi-set

$$C = \varphi\left[(A,a)\right] \cap \varphi\left[(B,b)\right] = \left(\varphi(A) \cap \varphi(B),\ \varphi \circ a \wedge \varphi \circ b\right). \tag{3.14}$$

(3.4) implies that $|C| \leq (1-\lambda)m$. An atom of measure β at point z equals $\pm|M-N|$ ("+" if $M > N$ and "$-$" otherwise). The contribution of that atom to the variation of β is given by

$$|M - N| = M + N - 2\varphi \circ a(z) \wedge \varphi \circ b(z). \tag{3.15}$$

Indeed, if $M > N$, then $\varphi \circ a(z) \wedge \varphi \circ b(z) = N$, and $|M-N| = M-N = M+N-2N$. The other case can be treated similarly. Using (3.12)–(3.15), we obtain

$$\|\beta\| = \sum_1^k r_i + \sum_1^\ell R_i - 2 \sum_{z \in \varphi(A) \cap \varphi(B)} \varphi \circ a(z) \wedge \varphi \circ b(z) = 2m - 2|C|. \tag{3.16}$$

(In fact, the argument shows that the number Θ in (3.8) for the measure α coincides with $|C|$.) From (3.4) it now follows that $\|\beta\| = \|\varphi \circ \alpha\| \geq \lambda 2m = \lambda\|\alpha\|$. Since measure $\mu = \dfrac{\alpha}{D}$, the required inequality $\|\varphi \circ \mu\| \geq \lambda\|\mu\|$ follows as well.

Let us show that $2 \Rightarrow 1$. Indeed, if there are given two multi-sets (A, a) and (B, b) with $|(A, a)| = |(B, b)| = m$, define measure α by assigning to each point $x_i \in A$ the atom $a(x_i) = r_i$, and at each point $y_i \in B$ the atom $b(y_i) = -R_i$. If φ is a function in F such that $\|\beta\| = \|\varphi \circ \alpha\| \geq \lambda\|\alpha\|$, then from (3.10) we obtain $|C| \leq (1-\lambda)m$, i.e., (3.4).

The equivalence $2 \Leftrightarrow 3$ follows from Lemma 3.4. Since all of the above classes S of measures contain the class $\ell^1_{0,K}$, uniform separation of measures of any such class S implies uniform separation of points, and, all the more so, strong separation of points. Now we can somewhat refine Theorems 2.5 and 2.6.

Uniform separation of points and discrete measures and superpositions of bounded functions. Let X, $X_1 \ldots$, X_N be sets, and let $\varphi_i : X \to X_i$, $i = 1, \ldots, N$, be mappings. The following theorem is a refinement of Theorem 2.6.

THEOREM 3.6. *The following conditions are equivalent.*

1. *An arbitrary function $f(x) \in B(X)$ can be represented in the form*

$$f(x) = g_1 \circ \varphi_1(x) + \cdots + g_N \circ \varphi_N(X), \qquad g_i \in B(X_i), \quad i = 1, \ldots, N. \tag{3.17}$$

2. *The family $F = \{\varphi_1, \ldots, \varphi_N\}$ uniformly separates points in X.*
3. *The family F uniformly separates measures of class $\ell^1(X)$.*
4. *The family F uniformly separates measures of class $\ell_0^1(X)$.*
5. *The family F uniformly separates measures of class $\ell_K^1(X)$.*
6. *The family F uniformly separates measures of class $\ell_{K,0}^1(X)$.*
7. *The family F uniformly separates measures of class $B(X)^\star$.*
8. *The family F uniformly separates measures of class $B_0(X)^\star$.*

PROOF. The equivalences $1 \Leftrightarrow 3 \Leftrightarrow 7$ are contained in Theorem 2.6. The equivalence $3 \Leftrightarrow 5$ follows from Lemma 3.4. Let us show that $1 \Leftrightarrow 8$. Let $\mathbb{R}^1$ be, as usual, the set of real numbers interpreted as a closed subspace in $B(X)$ or $B(X_i)$. Consider the quotient spaces

$$\widetilde{B}(X) = B(X)/\mathbb{R}^1, \qquad \widetilde{B}(X_i) = B(X_i)/\mathbb{R}^1.$$

Clearly, (3.17) is equivalent to the same equality understood in terms of equivalence classes from $B(X)$ and $\tilde{B}(X_i)$. As in Theorem 2.2 (and also 2.1), construct the space G only using $\tilde{B}(X_i)$ instead of $B(X_i)$. The dual space of G is $G^\star = B_0^\star(X_1) \times \cdots \times B_0^\star(X_N)$, where for $\nu = (\nu_1, \ldots, \nu_N)$, $\nu_i \in B_0^\star(X_i)$, we have $\|\nu\| = \max_{1 \le \nu \le N} \|\nu_i\|$. Then, following the scheme of the proof of Theorem 2.2, we find that (3.17) for equivalence classes is equivalent to existence of $\lambda > 0$ such that for an arbitrary $\mu \in B_0^\star(X)$

$$\max_{1 \le i \le N} \|\varphi_i \circ \mu\| \ge \lambda \|\mu\|,$$

i.e., $1 \Leftrightarrow 8$. Similarly, arguing as in the proof of Theorem 2.3 for spaces $\tilde{B}(X_i)$ we show that $1 \Leftrightarrow 4$. But from Lemma 3.4 it follows that $4 \Leftrightarrow 6$, whereas from Lemma 3.5 $4 \Leftrightarrow 2$. The proof is now complete.

Uniform separation of regular Borel measures and superpositions of continuous functions. In the case when X, X_i, $i = 1, \ldots, N$, are compact spaces and $\varphi_i : X \to X_i$ are continuous, we can slightly extend Theorem 2.5.

THEOREM 3.7. *The following are equivalent.*

1. *For all $f \in C(X)$ the following representation holds:*

$$f(x) = g_1 \circ \varphi_1(x) + \cdot + g_N \circ \varphi_N(X), \qquad g_i \in C(X_i). \tag{3.18}$$

2. *The family $F = \{\varphi_1, \ldots, \varphi_N\}$ uniformly separates regular Borel measures (i.e., measures of class $C(X)^\star$).*
3. *The family F uniformly separates measures of class $C(X)_0^\star$.*

PROOF. $1 \Leftrightarrow 2$ follows from Theorem 2.5. $1 \Leftrightarrow 3$ is seen by applying quotient spaces as in the previous theorem. One has to consider spaces $\tilde{C}(X_i) = C(X_i)/\mathbb{R}^1$ and argue according to the scheme in the proof of Theorem 2.1.

All the notions and results in this section are taken from Y. Sternfeld's papers [**133**] and [**134**].

§4. Constructing function families separating Borel measures

Let us first agree on some convenient terminology and notation.

(a) A family of subsets $\mathcal{U}$ of a topological space X is called *discrete* if all the closures of sets in $\mathcal{U}$ are pairwise disjoint.

(b) A function φ *separates* a family $\mathcal{U}$ of subsets of a topological space if for any two sets $V_1 \in \mathcal{U}$ and $V_2 \in \mathcal{U}$, $\varphi\left[\overline{V}_1\right] \cap \varphi\left[\overline{V}_2\right] = \emptyset$ (here, we are dealing with yet another type of separation).

(c) If $\mathcal{U}_1, \dots, \mathcal{U}_k$ are k families of subsets of a set X, we say that they *cover* X n times ($n \leq k$) if all $x \in X$ belong to some sets from families $\mathcal{U}_i$ for at least n indices i.

(d) If $\mathcal{U}$ is a family of subsets of a metric space X, then $\delta(\mathcal{U})$ denotes the least upper bound of diameters of sets in $\mathcal{U}$.

The following lemma is obvious.

LEMMA 4.1. *Let X be a set and $\{\mathcal{U}_i\}_{i=1}^k$ be k families of subsets of X. The following statements* (a), (b), (c), (d) *are equivalent, and all of them imply* (e).

(a) $\{\mathcal{U}_i\}_{i=1}^k$ *covers X n times.*

(b) *Any $k-n+1$ families out of $\{\mathcal{U}_i\}_{i=1}^k$ cover X one (sic!) time.*

(c) $\sum_{i=1}^k 1_{\mathcal{U}_i}(x) \geq n$ *for all $x \in X$, where* $1_E(x) = \begin{cases} 1, & x \in E, \\ 0, & x \in X \backslash E, \end{cases}$ *is the characteristic function of a set E.*

(From now on, we often shall not distinguish between a system $\mathcal{U}_i$ and a set of points from the union of all sets in $\mathcal{U}_i$.)

(d) *If all $\mathcal{U}_i$ belong to an algebra on which there is defined a probability measure μ, then*

$$\sum_{i=1}^k \mu(\mathcal{U}_i) \geq n.$$

(e) *Under the assumptions of* (d), *for any probability measure μ on X there exists i_0, $1 \leq i_0 \leq k$, such that $\mu(\mathcal{U}_{i_0}) \geq n/k$.*

The following lemma sheds some light on constructing systems of functions separating Borel measures.

LEMMA 4.2. *Let X be a compact metric space, and let $F = \{\varphi_i\}_1^k$ be a family of continuous functions on X. If for each $\varepsilon > 0$ there exist k finite discrete families $\mathcal{U}_1, \dots, \mathcal{U}_k$ of subsets of X such that*

$$\{\mathcal{U}_i\}_{i=1}^k \text{ covers } X \left[\tfrac{k}{2}\right]+1 \text{ times}, \tag{4.1}$$

$$\delta(\mathcal{U}_i) < \varepsilon, \qquad 1 \leq i \leq k; \tag{4.2}$$

$$\text{the } \varphi_i \text{ separate sets of the system } \mathcal{U}_i, \qquad 1 \leq i \leq k, \tag{4.3}$$

then the family F uniformly separates regular Borel measures on X (with constant $\lambda = 1/k$).

PROOF. We must show that for any $\mu \in C(X)^*$ there exists an index i, $1 \leq i \leq k$, such that $\|\varphi_i \circ \mu\| \geq (1/k)\|\mu\|$. Let $\mu = \mu^+ - \mu^-$ be the Jordan decomposition of the measure $\mu \in C(X)^*$. Since measures in $C(X)^*$ are regular, those measures for which the closed supports $S(\mu^+)$ and $S(\mu^-)$ of μ^+ and μ^- do not intersect form a dense (with respect to the norm) set in $C(X)^*$. In view of Lemma 3.4 it suffices to check our condition for those measures. Let μ be such a measure, and $\varepsilon > 0$ be the distance between $S(\mu^+)$ and $S(\mu^-)$. Assume $\|\mu\| = 1$, and hence $|\mu|$ (the variation of μ) is a probability measure. Consider the families $\mathcal{U}_i$, $i = 1, \ldots, k$, that form the covering guaranteed by the assumptions of the lemma with the given ε. Since $\delta(\mathcal{U}_i) < \varepsilon$, none of the sets in the system $\mathcal{U}_i$ can simultaneously intersect $S(\mu^+)$ and $S(\mu^-)$. In view of (4.1) and (e) of the previous lemma, there exists i_0, $1 \leq i_0 \leq k$, such that

$$|\mu|(\mathcal{U}_{i_0}) \geq \frac{1}{k}\left(\left[\frac{k}{2}\right] + 1\right) \geq \frac{1}{k}\left(\frac{k}{2} + \frac{1}{2}\right) = \frac{1}{2} + \frac{1}{2k}. \tag{4.4}$$

Since each of the sets in $\mathcal{U}_{i_0}$ can only intersect one of the sets $S(\mu^+)$ and $S(\mu^-)$ while φ_{i_0} separates sets in $\mathcal{U}_{i_0}$, we have

$$\left\|\varphi_{i_0}\left|_{\mathcal{U}_{i_0}} \circ \mu\right\| = |\mu|(\mathcal{U}_{i_0}) \geq \frac{1}{2} + \frac{1}{2k}. \tag{4.5}$$

(Here, $\varphi_{i_0}\left|_{\mathcal{U}_{i_0}}\right.$ is the restriction of φ_{i_0} to $\mathcal{U}_{i_0}$.) By (4.4),

$$|\mu|(X \backslash \mathcal{U}_{i_0}) \leq \frac{1}{2} - \frac{1}{2k}. \tag{4.6}$$

Therefore, in view of (4.5) and (4.6) we obtain

$$\|\varphi_{i_0} \circ \mu\| \geq \left\|\varphi_{i_0}\left|_{\mathcal{U}_{i_0}} \circ \mu\right\| - |\mu|(X \backslash \mathcal{U}_{i_0}) \geq \frac{1}{2} + \frac{1}{2k} - \left(\frac{1}{2} - \frac{1}{2k}\right) = \frac{1}{k}$$

and thus F uniformly separates Borel measures with constant $\lambda = 1/k$.

Ties to dimension. In view of Lemma 4.1 (for $n = [\frac{k}{2}] + 1$) each of $k - [\frac{k}{2}] = [\frac{k+1}{2}]$ families $\mathcal{U}_i$, $1 \leq i \leq k$, forms a covering of X. Naturally, for each of those coverings no more than $[\frac{k+1}{2}] \leq [\frac{k}{2}] + 1$ of the covering sets have a non-empty intersection (one from each family $\mathcal{U}_i$). So, the presence of the systems $\{\mathcal{U}_i\}_1^k$ with properties (4.1) and (4.2) implies that $\dim X \leq [\frac{k}{2}]$ (cf. [**72a**]). We shall make use of the following converse statement from dimension theory (Ostrand [**114**]).

LEMMA 4.3. *Let X be a compact metric space, $\dim X = n$ and $k \geq n + 1$, while $\varepsilon > 0$. There exist k finite discrete families $\{\mathcal{U}_i\}_1^k$ of subsets of X covering X $k - n$ times such that $\delta(\mathcal{U}_i) < \varepsilon$, $i = 1, \ldots, k$.*

Constructing separating families (continued).

THEOREM 4.4. *Let X be a compact metric space of dimension n. Quasi-all collections $(\varphi_1, \ldots, \varphi_{2n+1}) \in C(X)^{2n+1}$ uniformly separate Borel measures on X.*

PROOF. According to Lemma 4.3 construct the systems $\{\mathcal{U}_{i.m}\}$, $i = 1, \ldots,$ $2n+1$, $m = 1, 2, \ldots$, satisfying the following properties:

(a) Each $\mathcal{U}_{i,m}$ is a finite discrete family.

(b) For each m, the families $\{\mathcal{U}_{i,m}\}_{i=1}^{2n+1}$ cover X $n + 1 = \left[\dfrac{2n+1}{2}\right] + 1$ times.

(c) As $m \to \infty$, $\delta(\mathcal{U}_{i,m}) \to 0$, $i = 1, \ldots, 2n + 1$.

Denote by A_N a subset in $C(X)^{2n+1}$ consisting of the collections $(\varphi_1, \dots, \varphi_{2n+1})$ for which there exists $m \geq N$ such that $\{\varphi_i\}$ separate sets in the system $\mathcal{U}_{i,m}$, $i = 1, \dots, 2n+1$. Let us show that A_N is an open and everywhere dense set in $C(X)^{2n+1}$. Let $(\varphi_1, \dots, \varphi_{2n+1}) \in A_N$. For some $m \geq N$ the functions φ_i separate sets $\mathcal{U}_{i,m}$, $i = 1, \dots, 2n+1$. If ε_i is the minimal distance between $\varphi_i\left(\overline{U}\right)$ and $\varphi_i\left(\overline{V}\right)$, where $U \neq V$ run over $\mathcal{U}_{i,m}$, then in view of the discreteness of the system $\mathcal{U}_{i,m}$ we have $\varepsilon_i > 0$, and hence $\varepsilon = \min_i \varepsilon_i > 0$. If a collection $(\psi_1, \dots, \psi_{2n+1})$ from $C(X)^{2n+1}$ satisfies $\|\psi_i - \varphi_i\| < \frac{\varepsilon}{2}$, it is easily seen that the ψ_i separate $\mathcal{U}_{i,m}$. This proves that A_N is open in $C(X)^{2n+1}$.

Now, let $(\tau_1, \dots, \tau_{2n+1})$ be an arbitrary collection in $C(X)^{2n+1}$. We must show that arbitrarily close to it there exist collections from A_N. Since $\delta\left(\mathcal{U}_{i,m}\right) \to 0$, for all $\varepsilon > 0$ we can find $m \geq N$ such that the oscillations of functions τ_i on sets from $\mathcal{U}_{i,m}$ do not exceed $\varepsilon/2$.

Consider all the sets U_i^j that form the system $\mathcal{U}_{i,m}$, and for each of them define a rational number r_i^j such that $\|\tau_i - r_i^j\|_{U_i^j} < \varepsilon$. Take all the numbers r_i^j to be different. By the Tietze–Urysohn theorem construct a function $\varphi_i \in C(X)$ for which $\|\varphi_i - \tau_i\| < \varepsilon$, while $\varphi_i|_{U_i^j} = r_i^j$. Since all the numbers r_i^j are different, the function φ_i separates sets of the system $\mathcal{U}_{i,m}$. Repeating this construction for all i, we obtain a collection $(\varphi_1, \dots, \varphi_{2n+1}) \in A_N$ whose distance from $(\tau_1, \dots, \tau_{2n+1})$ is smaller than ε. The set $A := \bigcap_{N=1}^{\infty} A_N$ consists of quasi-all vectors in the space $C(X)^{2n+1}$. Each collection $(\varphi_1, \dots, \varphi_{2n+1}) \in A$ separates sets of systems $\mathcal{U}_{i,m}$ for an infinite set of indices m. Since $\delta\left(\mathcal{U}_{i,m}\right) \to 0$, according to Lemma 4.2 such a family $F = (\varphi_1, \dots, \varphi_{2n+1})$ uniformly separates Borel measures on X, and the theorem is proved.

A generalization of Kolmogorov's theorem. The Ostrand–Tikhomirov theorem. From the previous theorem and Theorem 3.7 there follows the following generalization of Kolmogorov's theorem, due to Tikhomirov [**87**] and Ostrand [**114**].

THEOREM 4.5. *Let X be a compact metric space of dimension n. For quasi-all collections $(\varphi_1, \dots, \varphi_{2n+1}) \in C(X)^{2n+1}$ the following holds: For any $f(x) \in C(X)$ there exist functions $g_1, \dots, g_{2n+1}$, $g_i \in C\left[\varphi_i(X)\right]$, such that*

$$f(x) = g_i\left(\varphi_1(X)\right) + \dots + g_{2n+1}\left(\varphi_{2n+1}(x)\right). \tag{4.7}$$

COROLLARY 4.6. *Under the assumptions of Theorem 4.5, for each $f(x) \in B(X)$ there exist functions $g_1, \dots, g_{2n+1}$, $g_i \in B\left[\varphi_i(X)\right]$, so that* (4.7) *holds.*

Further generalizations. Combining the above arguments, together with certain considerations used in the proof of Kolmogorov's theorem, one can obtain even more general results (Ostrand [**114**]).

LEMMA 4.7. *Let $X = X_1 \times \dots \times X_L$, where X_j, $j = 1, \dots, L$, are compact metric spaces. For each j, $1 \leq j \leq L$, let $\{\mathcal{U}_m^j\}$, $m = 1, 2, \dots$, be a sequence of finite discrete families of subsets of X_j and $\delta\left(\mathcal{U}_m^j\right) \to 0$ as $m \to \infty$. Let $\mathcal{U}_m$, $m = 1, 2, \dots$, be a family of subsets of X defined by*

$$\mathcal{U}_m = \left\{U^1 \times \dots \times U^L,\ U^j \in \mathcal{U}_m^j\right\}.$$

Finally, let $\lambda_1, \dots, \lambda_L$ *be real, rationally independent numbers. Quasi-all collections* $(\tau_1, \dots, \tau_L)$ *in* $C(X_1) \times \cdots \times C(X_L)$ *satisfy the following property: the functions*

$$\varphi(x_1, \dots, x_L) = \sum_{j=1}^{L} \lambda_j \tau_j(x_j)$$

separate sets of the system $\mathcal{U}_m$ *for infinitely many indices* m. *If* $X_1 = \cdots = X_L$ *and* $\mathcal{U}_m^1 = \cdots = \mathcal{U}_m^L$, *then for quasi-all functions* $\tau \in C(X_1)_L$ *the functions*

$$\varphi(x_1, \dots, x_L) = \sum_{j=1}^{L} \lambda_j \tau(x_j)$$

separate sets $\mathcal{U}_m$ *for infinitely many indices* m.

THEOREM 4.8. *Let* $X = X_1 \times \cdots \times X_L$, *where* X_j *are compact metric spaces of dimension* n_j, $j = 1, \dots, L$. *Set* $n = \sum_1^L n_j$. *There exist functions* $\psi_{i,j} \in C(X_j)$, $i = 1, \dots, 2n+1$, $j = 1, \dots, L$, *so that any function* $f \in C(X)$ *can be represented in the form*

$$f(x_1, \dots, x_L) = \sum_{i=1}^{2n+1} g_i \left(\sum_{j=1}^{L} \psi_{i,j}(x_j) \right), \quad \text{where } g_i \in C(\mathbb{R}). \tag{4.8}$$

If $X_1 = X_2 = \cdots = X_L$ *and* $\lambda_1, \cdots, \lambda_L$ *are rationally-independent numbers, then* $\psi_{i,j}$ *can be chosen to be* $\lambda_j \varphi_i$, $\varphi_i \in C(X_1)$, *and* $n = n_1 L$. *For* $f(x_1, \dots, x_L) \in C(X_1^L)$, *we then have the representation*

$$f(x_1, \dots, x_L) = \sum_{1}^{2n+1} g_i \left(\sum_{j=1}^{L} \lambda_j \varphi_i(x_j) \right). \tag{4.9}$$

Setting $X_1 = I$, $n_1 = 1$, $n = L$, we obtain from (4.9) Kolmogorov's formula (1.1), although, generally speaking, all the g_i may now be different.

To prove Theorem 4.8, one has to combine Lemma 4.3 and Theorem 4.4 (the latter must be applied several times). We omit the details.

COROLLARY 4.9. *The last theorem holds if we consider* $f(x_1, \dots, x_L) \in B(X)$ *and* $g_i \in B(\mathbb{R})$.

The entire contents of this section is taken from Sternfeld's paper [**134**].

§5. Dimension and the number of terms in the Kolmogorov representation

The following natural question arises in connection with the theorems of Kolmogorov, Ostrand, and Tikhomirov. Is it possible in the representation of an arbitrary function $f(x) \in C(X)$ on an n-dimensional compact space X by the formula

$$f(x) = \sum_{1}^{N} g_i(\varphi_i(x)), \tag{5.1}$$

where $\varphi_i(x) \in C(X)$ and $g_i \in C(\varphi_i(X))$, to have the number of terms $N < 2n+1$? For a long time only two partial results have been known, relating to functions of two variables on the square I^2. In [**39**], Doss proved that for arbitrary monotone continuous functions $\varphi_{pq}(x)$, $p = 1, 2$, $q = 1, \dots, 4$, there exists a function $f(x_1, x_2) \in C(I^2)$ that cannot be represented in the form

$$f(x_1, x_2) = \sum_{q=1}^{4} g_q \left[\varphi_{1q}(x_1) + \varphi_{2q}(x_2)\right], \qquad g_q \in C(\mathbb{R}).$$

Bassalygo [**22**] showed that for every collection $\{\varphi_i(x_1, x_2)\} \subset C(I^2)$, $i = 1, 2, 3$, there exists $f(x_1, x_2) \in C(I^2)$ that cannot be represented by (5.1) with $N = 3$.

The possibility of representation (5.1) implies, in particular, that the collection of functions $(\varphi_1(x), \dots, \varphi_N(x))$ defines a homeomorphic embedding of the n-dimensional compact space X into $\mathbb{R}^N$. The possibility of such an embedding of an n-dimensional X into $\mathbb{R}^{2n+1}$ is an old result of dimension theory: the Menger–Nöbeling theorem ([**72a,** p.89]). However, for a particular compact space, the dimension $2n+1$ can sometimes be lowered. For example, I^n can be embedded into $\mathbb{R}^n$ by the identity map. At the same time, functions $\varphi_i(x)$ providing the representation (5.1) not only separate points on X (which is necessary for the embedding to be a homeomorphism), but also, according to Theorem 3.7, uniformly separate Borel measures. As it turns out, for such systems the number $2n+1$ clearly becomes a rigid characteristic of the dimension n.

THEOREM 5.1. *If* $\dim X = n$ $(n \geq 1)$ *and a system of functions* $\varphi_i(x) \in C(X)$, $i = 1, \dots, N$, *uniformly separates Borel measures on* X, *then* $N \geq 2n+1$, *and there exists a system of* $(2n+1)$ *functions that does so.*

Thus, for the representation (5.1) to hold for all $f(x) \in C(X)$ when $\dim X = n$, it is necessary that $N \geq 2n+1$. Theorem 5.1 (its first part—the second part is the theorem of Ostrand and Tikhomirov) was proved by Sternfeld. First, in [**132**], Bassalygo's result was improved to $N = 4$; later in [**132**] the theorem was proved for $n = 2, 3, 4$; and finally, in [**135**] it was established in full generality by a rather tedious argument.

Even a stronger result holds:

THEOREM 5.2. *If* $\dim X = n \geq 2$, *then any system of functions* $\varphi_i(x) \in C(X)$, $i = 1, \dots, N$, *uniformly separating points in* X *contains at least* $2n+1$ *functions*: $N \geq 2n+1$.

Thus, the condition $\dim X = n$ is characterized by the fact that in (5.1), or in a similar representation for an arbitrary function $f(x) \in B(X)$ and with $g_i \in B(\mathbb{R})$, there are at least $2n+1$ terms. This is a complete solution of the questions about the possibility of decreasing the number of terms in the Kolmogorov–Ostrand–Tikhomirov representation (and, moreover, obtained without appealing to any special structure of inner functions in Kolmogorov's theorem for $X = I^n$).

The dimension of X and the structure of $C(X)$. For a fixed $\varphi(x) \in C(X)$, consider the set $A = \{g(\varphi(x)), g \in C(\mathbb{R})\}$. A is a closed subalgebra of $C(X)$ that contains constants and is generated by one element $\varphi(x)$. Conversely, any closed subalgebra of $C(X)$ generated by one element $\varphi(x)$ and containing constants is of the form $\{g(\varphi(x)), g \in C(\mathbb{R})\}$. Theorem 5.1 can be reformulated in the following way.

THEOREM 5.3. $\dim X = n$ *if and only if* $C(X)$ *can be represented by an algebraic sum of* $2n+1$ *closed subalgebras, each of which is generated by one element and cannot be represented by such a sum with less than* $2n+1$ *subalgebras.*

§6. Other types of separation

Which notion of separation of measures or points corresponds to those theorems on representations by superpositions in which, as in Theorems 1.1 and 1.2, the outer function is the same for all terms? Let X, X_i, $i = 1, \dots, N$, be compact spaces, let $\varphi_i : X \to X_i$ be continuous mappings. Suppose that the topologies in X_i are coordinated in the following sense: If $X_i \cap X_j \neq \emptyset$, then the topologies induced in $X_i \cap X_j$ by those in X_i and X_j coincide. Set $Z = \bigcup_{j=1}^{N} X_j$ with the natural topology on Z defined by taking the union of the topologies on the X_j. To each $g \in C(Z)$ we relate a function $f \in C(X)$ by the formula

$$f(x) = \sum_{1}^{N} g\left(\varphi_i(x)\right). \tag{6.1}$$

THEOREM 6.1. *In order that for all* $f \in C(X)$ *representation* (6.1) *hold with some* $g \in C(Z)$, *it is necessary and sufficient that there exists a constant* λ, $0 < \lambda \leq N$, *such that for any* $\mu \in C(X)^\star$

$$\|\nu\| \geq \lambda \|\mu\|, \quad \textit{where } \nu = \sum_{1}^{N} \varphi_i \circ \mu \qquad (\nu \in C(Z)^\star). \tag{6.2}$$

PROOF. Consider an operator $A : C(Z) \to C(X)$ that associates to each $g(z) \in C(Z)$ the function $f(x) \in C(X)$ by formula (6.1). The adjoint operator $A^\star : C(X)^\star \to C(Z)^\star$ acts by the formula $\mu \in C(X)^\star \mapsto \nu = \sum_{1}^{N} \varphi_i \circ \mu$. This can be checked by direct arguments similar to those in Theorems 3.6 and 3.7. Thus, the inequality (6.2) is a sufficient condition for surjectivity of the operator A. That λ cannot be larger than N follows directly from the inequality $\|\varphi_i \circ \mu\| \leq \|\mu\|$.

REMARK. The reader may be somewhat puzzled by the fact that now $\lambda \leq N$, whereas in Theorems 3.6 and 3.7 it was $\lambda \leq 1$. Here is a simple explanation. For the sake of clarity, let $X_1 = \cdots = X_N = Z$. In Theorems 3.6 and 3.7 we took $C = C(X_1) \times \cdots \times C(X_N)$, while for $g = (g_1, \dots, g_N) \in C$, $\|g\|$ means $\sum_{1}^{N} \|g_i\|$, and accordingly, in $C^\star$ for $\nu = (\nu_1, \dots, \nu_N)$ we had $\|\nu\| = \max_{1 \leq i \leq N} \|\nu_i\|$. So, for $\mu \in C(X)^\star$ and $\nu = A^\star \mu = (\varphi_1 \circ \mu, \dots, \varphi_N \circ \mu) \in C^\star$ we had the inequality $\|\nu\| = \max_i \|\varphi_i \circ \mu\| \leq \|\mu\|$, and hence λ could not exceed 1. This also meant that for $\tilde{g} \in C$, $\|\tilde{g}\| = \sum_{1}^{N} \|g_i\| \leq 1$, we had

$$|\langle A\tilde{g}, \mu\rangle| = |\langle \tilde{g}, A^\star \mu\rangle| = \left| \sum_{1}^{N} \int_{X_i} g_i d\left(\varphi_i \circ \mu\right) \right| \leq \max_{1 \leq i \leq N} \|\varphi_i \circ \mu\| \leq \|\mu\|.$$

Now, if all $g_i = g$, $i = 1, \dots, N$, and still $\|\tilde{g}\| = N\|g\| \le 1$, the latter inequality implies $|\langle \tilde{g}, \nu \rangle| \le \|\mu\|$. However, if $\|g\| \le 1$, as in Theorem 6.1, this last estimate increases by a factor of N.

Of course, inequality (6.2) implies to some extent a condition of separation of measures that is naturally stronger than uniform separation. This separation can be characterized in a way similar to Lemmas 3.1–3.3. For a measure μ let $X = S^+ \cup S^-$, $S^+ \cap S^- = \emptyset$, and for each $E \subset S^+$ let $\mu(E) > 0$, while $\mu(E) < 0$ for each $E \subset S^-$ (S^+, S^- is the Hahn decomposition for the measure μ). Let $P \subset Z$ and $E_i^+ = \varphi_i^{-1}(P) \cap S^+$, $E_i^- = \varphi_i^{-1}(P) \cap S^-$. For the measure $\nu = \sum_1^{N'} \varphi_i \circ \mu$ we have

$$\nu(P) = \sum_1^N \mu\left(E_i^+\right) + \mu\left(E_i^-\right).$$

Set

$$\Theta(P) = \min\left(\sum_1^N \mu\left(E_i^+\right), \sum_1^N \left|\mu\left(E_i^-\right)\right|\right). \tag{6.3}$$

Then,

$$|\nu(P)| = \sum_{j=1}^N \mu\left(E_i^+\right) + \sum_{j=1}^N \left|\mu\left(E_i^-\right)\right| - 2\Theta(P). \tag{6.4}$$

Consider a partition of $Z : Z = \bigcup_j P^j$, and let

$$E_i^{j+} = \varphi_i^{-1}\left(P^j\right) \cap S^+, \quad E_i^{j-} = \varphi_i^{-1}\left(P^j\right) \cap S^-. \tag{6.5}$$

Then for a fixed i, the sets E_i^{j+} form a partition of S^+, while the E_i^{j-} form one for S^-. Finally, define the number Θ by

$$\Theta = \Theta(\mu) = \inf \sum_j \Theta\left(P^j\right), \tag{6.6}$$

where the infimum is taken over all partitions of Z. The next statement follows immediately.

LEMMA 6.2. *If a measure* $\nu = \sum_1^N \varphi_i \circ \mu$, *then*

$$\|\nu\| = N\|\mu\| - 2\Theta(\mu). \tag{6.7}$$

PROOF. Indeed, for a given partition Z into sets P^j we have

$$\begin{aligned}\sum_j |\nu(P^j)| &= \sum_j \left[\sum_{i=1}^N \mu\left(E_i^{j+}\right) + \sum_{i=1}^N \left|\mu\left(E_i^{j-}\right)\right| - 2\Theta\left(P^j\right)\right] \\ &= \sum_{i=1}^N \sum_j \left[\mu\left(E_i^{j+}\right) + \left|\mu\left(E_i^{j-}\right)\right|\right] - 2\sum_j \Theta\left(P^j\right) \\ &= N\left[\mu\left(S^+\right) + \left|\mu\left(S^-\right)\right|\right] - 2\sum_j \Theta\left(P^j\right) = N\|\mu\| - 2\sum_j \Theta\left(P^j\right).\end{aligned}$$

Taking the supremum over all portions of Z, we obtain (6.7).

An immediate corollary is the following lemma.

LEMMA 6.3. *The separation condition* (6.2) *is equivalent to the following inequality holding for all* $\mu \in C(X)^\star$:

$$2\Theta(\mu) \leq (N - \lambda)\|\mu\|.$$

Now, let X, X_i, $i = 1, \dots, N$, be arbitrary sets, and let $\varphi_i : X \to X_i$ be mappings. Let $Z = \bigcup_{i=1}^N X_i$. Arguing as above, we obtain the following result.

THEOREM 6.4. *In order that for all* $f \in B(X)$ *the representation* (2.39) *hold with some* $g \in B(Z)$, *it is necessary and sufficient that there exists a constant* λ, $0 < \lambda \leq N$, *such that for all* $\mu \in \ell^1(X)$

$$\|\nu\| \geq \lambda\|\mu\|, \quad \textit{where } \nu = \sum_1^N \varphi_i \circ \mu \qquad (\nu \in B(Z)^\star). \tag{6.8}$$

The inequality (6.8) *is equivalent to* (6.2) *for* $\mu \in \ell^1(X)$.

COROLLARY 6.5. *If, under the assumptions of Theorem* 6.1, *for each* $f(x) \in C(X)$ *the representation* (6.1) *holds with some* $g \in C(Z)$, *then for any* $f(x) \in B(X)$ *the representation* (6.1) *also holds with some* $g \in B(Z)$.

COROLLARY 6.6. *Under the assumptions of Kolmogorov's theorem, for each* $f(x) \in B(I^n)$ *the representation* (1.1) *holds with some* $g \in B(I)$.

(Corollary 6.6 slightly sharpens Corollary 2.8.)

More on separation. Again, let X, X_i be compact spaces, $\varphi_i : X \to X_i$, $i = 1, \dots, N$, continuous mappings. Let D_i be a Banach space, $D_i \subset C(X_i)$ (the inclusion is understood in the set-theoretic sense only). We also assume that there exist constants $d_i > 0$ such that for $g \in D_i$ we have the inequality $\|g\|_{C(X_i)} \leq d_i\|g\|_{D_i}$. This implies, in particular, that any linear functional continuous on D_i with respect to the norm on $C(X_i)$ is also continuous on D_i with respect to its own norm. Thus, a regular Borel measure ν on X_i defines on D_i a continuous functional, whose norm on D_i we shall denote by $\|\nu\|_{D_i}$. Clearly, it is possible that $\|\nu\| > 0$ while $\|\nu\|_{D_i} = 0$.

Consider the natural question of whether it is possible to represent an arbitrary $f(x) \in C(X)$ in the form

$$f(x) = \sum_1^N g_i(\varphi_i(x)), \qquad g_i \in D_i. \tag{6.9}$$

THEOREM 6.7. *In order that an arbitrary $f(x) \in C(X)$ be representable by* (6.9), *it is necessary and sufficient that there exists $\lambda > 0$ such that for any $\mu \in C(X)^\star$ and some index i, $1 \le i \le N$,*

$$\|\varphi_i \circ \mu\|_{D_i} \ge \lambda \|\mu\|. \tag{6.10}$$

Now consider the situation in Theorem 6.1, let $D \subset C(Z)$ be a Banach space, and let there exist $d > 0$ such that for $g \in D$, $\|g\|_{C(Z)} \le d\|g\|_D$. A regular Borel measure ν on Z defines a continuous linear functional on D, whose norm we denote by $\|\nu\|_D$.

THEOREM 6.8. *In order that in the situation of Theorem* 6.1 *the representation* (6.1) *for arbitrary $f \in C(X)$ hold with $g \in D$, it is necessary and sufficient that there exists $\lambda > 0$ such that for any $\mu \in C(X)^\star$,*

$$\left\| \sum_1^N \varphi_i \circ \mu \right\| \ge \lambda \|\mu\|. \tag{6.11}$$

Consider, as in Theorem 1.2, the subspace D in the space $C(I)$ that consists of boundary values $g\left(e^{ix}\right)$, $x \in I$, of functions $g(z)$ analytic for $|z| < 1$ and continuous for $|z| \le 1$. We shall not specify here the precise form of the condition (6.11) in these circumstances, but simply note that calculation of the norm of the functional ν on such D is itself a popular extremal problem in the theory of analytic functions. (Note that the results 6.1–6.8 appear here for the first time.)

Superpositions of functions of one variable. Theorems 6.7 and 6.8 are not only associated with superpositions of functions of several variables, but can also be useful for functions of one variable when the question concerns representation of functions by superpositions of functions having some "good" additional properties. Let us give here a well-known example. Let $X = \mathbb{T}$ be the unit circle and $D = A(\mathbb{T}) \subset C(\mathbb{T})$ be the Banach space of absolutely convergent Fourier series. According to a theorem of Kahane ([**73**, p.122]) an arbitrary $f \subset C(\mathbb{T})$ can be represented in the form

$$f = g \circ \varphi_1 + g \circ \varphi_2 + g \circ \varphi_3, \tag{6.12}$$

where $g \in A(\mathbb{T})$ and φ_1, φ_2, φ_3 are self-homeomorphisms of $\mathbb{T}$ satisfying some additional continuity properties.

Thus such triples $(\varphi_1, \varphi_2, \varphi_3)$ provide some kind of separation as in (6.11). If we take pairs (φ_1, φ_2) of conjugate homeomorphisms, then they only provide representations

$$f = g_1 \circ \varphi_1 + g_2 \circ \varphi_2$$

of an arbitrary $f \in C(\mathbb{T})$, and therefore such pairs (φ_1, φ_2) only provide separation (6.10). In addition, since the set of functions $g \circ \varphi_1 + g \circ \varphi_2$ does not coincide with $C(\mathbb{T})$ (cf. [**73**, p.124]), the pair (φ_1, φ_2) does not provide separation as in (6.11). Hence, conditions (6.10) and (6.11) are indeed different.

§7. Study of the original notion of separation (continued)

Let us go back to the original definition of uniform separation of measures. Let X, X_i, $i = 1, \dots, N$, be arbitrary sets and $\varphi_i : X \to X_i$ be mappings. We continue the study of separating properties of the family $F = \{\varphi_i\}_1^N$. Let $(X, \mathcal{M})$ be a measurable space, and let S be a class of real-valued measures defined on $\mathcal{M}$. Let $|Z|$ denote the cardinality of a set Z, and for sets $Z \subset X$ define the following derivatives:

$$Z^i = \left\{x \in Z : \left|Z \cap \varphi_i^{-1}(\varphi_i(x))\right| \geq 2\right\}; \tag{7.1}$$

$$\tau(Z) = \bigcap_{i=1}^{N} Z^i. \tag{7.2}$$

Thus, Z^i is the set of all points in Z for each of which there is at least one more point where the function φ_i assumes the same value. Operator $\tau : Z \to \tau(Z)$ acts in the space of all subsets of the set X. The following sufficient condition for uniform separation of measures then holds.

LEMMA 7.1 ([**133**]). *Let mappings $\varphi_i \in F$ be such that for $Z \in \mathcal{M}$ the sets Z^i (as in* (7.1)*) also belong to $\mathcal{M}$, $i = 1, \dots, N$. If $\tau^n(X) = \emptyset$ for some $n \geq 1$, then F uniformly separates measures in S.*

PROOF. Use induction on n. Clearly, for $n = 0$ we can assume that the statement holds. Suppose it holds for $n-1$ for measures supported on arbitrary subsets in $\mathcal{M}$. Let $\tau^n(X) = \emptyset$ and set $Z = \tau(X)$. Then $\tau^{n-1}(Z) = \tau^n(X) = \emptyset$. Hence, measures in S concentrated on Z are uniformly separated by the family F. Suppose this separation occurs with a constant λ, $0 < \lambda \leq 1$. Take a number α, $\dfrac{1}{1+\lambda} < \alpha < 1$, and let the measure $\mu \in S$, $\|\mu\| = 1$. By $\mu|_Z$ we denote the restriction of μ to the set Z, and by $|\mu|$ the total variation of μ. Consider two cases.

1. $|\mu|(Z) \geq \alpha$. So, $\|\mu|_Z\| \geq \alpha$ and there exists i such that $\|\varphi_i \circ \mu|_Z\| \geq \alpha\lambda$. Since $|\mu|(X \backslash Z) \leq 1 - \alpha$,

$$\|\varphi_i \circ \mu\| \geq \|\varphi_i \circ \mu|_Z\| - |\mu|(X \backslash Z) \geq \lambda\alpha - (1-\alpha) = (\lambda+1)\alpha - 1 > 0.$$

2. $|\mu|(Z) < \alpha$. Then, $|\mu|(X \backslash Z) > 1 - \alpha$. Recalling that $Z = \tau(X)$, we have

$$X \backslash Z = X \backslash \bigcap_{i=1}^{N} X^i = \bigcup_{i=1}^{N} \left(X \backslash X^i\right).$$

Therefore, for some i we must have $|\mu|\left(X \backslash X^i\right) > \dfrac{1-\alpha}{N}$. Now, note that $\varphi_i\left(X^i\right) \cap \varphi_i\left(X \backslash X^i\right) = \emptyset$. Indeed, if $y \in \varphi_i\left(X^i\right)$, then y has at least two preimages, while $y \in \varphi_i\left(X \backslash X^i\right)$ has only one preimage. So, on $X \backslash S^i$ the mapping is a bijection. Hence,

$$\|\varphi_i \circ \mu\| \geq \|\varphi_i \circ \mu\|_{\varphi_i(X \backslash X^i)} = \|\mu\|_{X \backslash X^i} = |\mu|\left(X \backslash X^i\right) > \frac{1-\alpha}{N}.$$

So, if $\lambda' = \min((1+\lambda)\alpha - 1, (1-\alpha)/N)$, then, given that $\|\mu\| = 1$, we obtain that there exists an index i for which $\|\varphi_i \circ \mu\| \geq \lambda'$. Thus, F uniformly separates measures on S (with constant λ').

THEOREM 7.2 ([**133**]). *If $F = \{\varphi_1, \varphi_2\}$ consists of two functions, then this family uniformly separates points in X if and only if $\tau^n(X) = \emptyset$ for some n.*

PROOF. We can assume now that $\mathcal{M}$ consists of all subsets of X while $S = \ell^1(X)$. Sufficiency of the condition $\tau^n(X) \neq \emptyset$ has been established in the above lemma. It remains to show necessity. Thus, let $\tau^n(X) \neq \emptyset$ for all n. Let $x_1 \in \tau^n(X) = \left[\tau^{n-1}(X)\right]^1 \cap \left[\tau^{n-1}(X)\right]^2$. Since $x_1 \in \left[\tau^{n-1}(X)\right]$, according to (7.1) there exists $x_2 \in \tau^{n-1}(X)$ for which $\varphi_1(x_2) = \varphi_1(x_1)$. Since $x_2 \in \tau^{n-1}(X) = \left[\tau^{n-2}(X)\right]^1 \cap \left[\tau^{n-2}(X)\right]^2$, there exists an element $x_3 \in \tau^{n-2}(X)$ such that $\varphi_2(x_3) = \varphi_2(x_2)$. Since $x_3 \in \tau^{n-2}(X) = \left[\tau^{n-3}(X)\right]^1 \cap \left[\tau^{n-3}(X)\right]^2$, there exists $x_4 \in \tau^{n-3}(X)$ such that $\varphi_1(x_4) = \varphi_1(x_3)$. Continuing, we obtain a sequence $x_1, \dots, x_n$, $x_j \in \tau^{n-j+1}(X)$, and $\varphi_1(x_j) = \varphi_1(x_{j+1})$ for odd j, while $\varphi_2(x_j) = \varphi_2(x_{j+1})$ for even j. Set

$$\mu = \frac{1}{n} \sum_1^n (-1)^j \delta_{x_j}$$

(δ_x denotes the delta-mass at a point x). Then $\|\mu\| = n$, and at the same time $\|\varphi_i \circ \mu\| \leq 2/n$, $i = 1, 2$. Thus, $\|\varphi_i \circ \mu\| \leq \dfrac{2}{n}\|\mu\|$, and since n can be arbitrarily large, it follows that the family $\{\varphi_1, \varphi_2\}$ does not uniformly separate points in X.

We mention in passing that sequences of points similar to that constructed in the proof of Theorem 7.2 will play quite an important role in the sequel.

Theorem 7.2 together with Theorem 3.6 provides a complete characterization of sets on which an arbitrary bounded function is representable by a sum of two superpositions.

THEOREM 7.3. *Let X, X_1, X_2 be sets, and $\varphi_i : X \to X_i$ mappings. In order that any function in $B(X)$ be representable in the form*

$$f(x) = g_1 \circ \varphi_1(x) + g_2 \circ \varphi_2(x), \qquad g_i \in B(X_i), \quad i = 1, 2, \tag{7.3}$$

it is necessary and sufficient that $\tau^n(X) = \emptyset$ for some n.

A question and a theorem of Sternfeld. Now we are able to give a complete answer to the earlier question on equivalence of representability of continuous functions by superpositions of continuous functions and its analogue for bounded functions (the question was raised in [**133**]).

THEOREM 7.4 ([**133**]). *Let X, X_i be compact metric spaces, and $\varphi_i : X \to X_i$ be continuous mappings, $i = 1, 2$. The following statements are equivalent.*

1. *For any $f \in C(X)$ the following representation holds:*

$$f(x) = g_1 \circ \varphi_1(x) + g_2 \circ \varphi_2(x), \qquad g_i \in C(X_i), \quad i = 1, 2. \tag{7.4}$$

2. *For any $f(x) \in B(X)$, (7.3) holds.*
3. *The system $F = \{\varphi_1, \varphi_2\}$ uniformly separates Borel measures on X.*
4. *The system $F = \{\varphi_1, \varphi_2\}$ uniformly separates points in X.*
5. *The equality $\tau^n(X) = \emptyset$ holds for some n.*

PROOF. First, we shall need the following technical lemma. Here $\mathcal{M}$ is an algebra of Borel sets.

LEMMA 7.5. *If $Z \in \mathcal{M}$, then Z^i, $i = 1, 2$, and $\tau(Z)$, defined by (7.1)–(7.2), also belong to $\mathcal{M}$.*

PROOF OF THE LEMMA. Let $d(U)$ denote the diameter of a set U. We have

$$\begin{aligned} Z^i &= \left\{x \in Z : \left|Z \cap \varphi_i^{-1}\left(\varphi_i(x)\right)\right| \geq 2\right\} \\ &= \left\{x \in Z : d\left(Z \cap \varphi_i^{-1}\left(\varphi_i(x)\right)\right) > 0\right\} \\ &= \bigcup_{n=1}^{\infty} \left\{x \in Z : d\left(Z \cap \varphi_i^{-1}\left(\varphi_i(x)\right)\right) \geq \frac{1}{n}\right\}. \end{aligned}$$

Since $\varphi_i^{-1}(\varphi_i(x))$ is compact, $\{x \in Z : d(Z \cap \varphi_i^{-1}(\varphi_i(x)) \geq 1/n)$ is closed, and hence Z^i is an F_σ set in Z. Therefore, Z^i is a Borel set.

PROOF OF THEOREM 7.4. We already know from the preceding sections that $1 \Leftrightarrow 3$, $2 \Leftrightarrow 4 \Leftrightarrow 5$, and also $1 \Rightarrow 2$. Now suppose 2 holds. Then 5 also holds. But from Lemma 7.1, together with the above Lemma 7.5, it follows that 3 holds. The theorem is proved.

Theorem 7.4 was independently (and in a different form) proved by Khavinson [**80**] using different arguments that allow us to remove the assumption that the compact sets are metric. We shall present that argument later on.

§8. A counterexample

It is very surprising that for $N > 2$ Theorem 7.4 is false. Examples were given in [**136**] and [**103**]. Here, we follow [**103**]. Yet, we have to start out from afar.

A free group. Let us be given a countable set of symbols $v_1, \ldots, v_n, \ldots$. Add to them the symbols $v_1^{-1}, \ldots, v_n^{-1}, \ldots$ and a symbol e_0 that will play the role of the identity. From this alphabet we look for words, i.e., symbols, that have the form

$$F = f_m f_{m-1} \cdots f_1, \tag{8.1}$$

where the f_i are arbitrary symbols from the alphabet, and $m \in \mathbb{N}$ is arbitrary. Moreover, nowhere in (8.1) do symbols v_n and v_n^{-1}, for any $n \in \mathbb{N}$, stand near each other. We postulate $v_n v_n^{-1} = v_n^{-1} v_n = e_0$ ($v_n v_n^{-1}$ or $v_n^{-1} v_n$ are "empty" words). Multiplication of words that is defined accordingly (by writing them in order with the possible omission of "empty" words) makes the set of all words G_0 into a group with e_0 serving as its identity. If in the word (8.1) several consecutive symbols coincide, we replace them by a power, as usual. The group G_0 is called a free group with generators $v_1, \ldots, v_n, \ldots$. The subgroup of G_0 with generators $v_1, \ldots, v_n$ will be denoted by G_n; it is a free group with a finite alphabet $\left\{v_1, v_1^{-1}, \ldots, v_n, v_n^{-1}, e_0\right\}$.

An isomorphism of a free group into a group of analytic homeomorphism of the semi-axis $[0, +\infty)$**.** Denote by G the set of functions $v(x)$ mapping the semi-axis $[0, +\infty)$ onto itself and satisfying the following properties:

$$\begin{gathered} v(0) = 0, \qquad v'(x) > 0 \quad \text{for all } x > 0, \qquad v'(0) = 1, \\ v(x) \text{ is an analytic function on } [0, +\infty). \end{gathered} \tag{8.2}$$

Clearly, for $v(x) \in G$, $v^{-1}(x) \in G$. The set G becomes a group if for group multiplication of $u \in G$ and $v \in G$ we take the superposition $u \circ v$. The identity map $e : e(x) \equiv x$ plays the role of the identity. Right away, let us note an important (for us) example of functions in G. Let $[a,b] \subseteq [0,+\infty)$, and let $\mu \geq 0$ be a regular Borel measure on $[a,b]$ satisfying

$$\int_a^b d\mu = 1. \tag{8.3}$$

The function

$$v(x) = x \int_a^b e^{xt} d\mu_t \tag{8.4}$$

is an entire analytic function and belongs to G.

Choose in G certain functions $v_1(x), \ldots, v_n(x)$, and define a homomorphism φ of a free group G_0 into the group G by setting

$$\varphi(v_i) = v_i(x), \qquad \varphi(e_0) = e. \tag{8.5}$$

Clearly, φ extends naturally to all words (8.1) and is indeed a homomorphism of a free group G_0 into G. If F is a word (8.1) and φF its image, we shall often write $F(x)$ instead of $(\varphi F)(x)$, since the presence of the variable x already distinguishes a function (an element of G) from a word.

The following fact, also established by Medvedev, plays an important role in his construction.

THEOREM 8.1. *Functions $v_1(x), \ldots, v_n(x)$ in the group G can be chosen so that the mapping* (8.5) *is an isomorphism of G_0 onto $\varphi(G_0)$. Moreover, the functions $v_n(x)$ also satisfy the following additional properties*:

$$0 < v_{n+1}^{(k)}(x) < v_n^{(k)}(x), \tag{8.6}$$

for all $k = 0, 1, \ldots$ and all n.

Auxiliary facts from the problem of moments. The proof of Theorem 8.1 is based on some facts from the problem of moments which we gather in the following lemma. Let $0 < x_1 < x_2 < \cdots < x_n$ be certain points, and $c_1, \ldots, c_n$ real numbers. Let

$$\Omega\begin{pmatrix} x_1, & \ldots, & x_n \\ c_1, & \ldots, & c_n \end{pmatrix}$$

denote the class of functions $v(x)$ of the type (8.4) for which the measure $\mu \geq 0$ has infinitely many points of growth, and the interpolation conditions $v(x_i) = c_i$, $i = 1, \ldots n$, are satisfied. (We denote by $\mathcal{P}$ the class of measures $\mu \geq 0$ on $[a,b]$ having infinitely many points of growth and normalized by (8.3).)

LEMMA 8.2. *Let the class Ω be non-empty. If a point x is different from $x_1, \ldots, x_n$, then the set $S(x) = \{v(x) : v \in \Omega\}$ is a non-degenerate open interval. If $\xi \in S(x)$, then $\xi \in S(y)$ for all y sufficiently close to x.*

Both statements of the lemma follow easily from standard facts of the theory of moments. In particular, one could, with this goal in mind, compare Theorems 3.5, 6.1, and 4.1 in [**89**].

PROOF OF THEOREM 8.1. First, let $v_1(x), \dots, v_n(x), \dots$ be an arbitrary sequence of functions in G. Let us sort out the structure of words F inside the kernel of the homomorphism φ. If for a word $F = f_m \dots f_1$ we have $\varphi F = e$, then for the word $F_1 = f_1 f_m \dots f_2$ we have $\varphi(F_1) = \varphi(f_1)\varphi(F)\varphi(f_1^{-1}) = \varphi(f_1) e \varphi(f_1^{-1}) = \varphi(f_1 f_1^{-1}) = e$. Hence, if $\varphi(F) = e$, then for all words F_1 obtained from F by a circular permutation of the symbols we have $\varphi(F_1) = e$. Thus, if the end-terms f_1 and f_m of a word are reciprocal of one another, they can be deleted. If the new end-terms of the word are reciprocal, we again can delete them, etc. All the terms in F will not be deleted; otherwise, the neighboring terms in the middle would be mutually reciprocal, and this is not allowed for words.

Mark some $n \in \mathbb{N}$. By circular permutations and cancellations, the equality $\varphi(F) = e$ can be reduced to the equivalent equality $\varphi(F_0) = e$, where F_0 has one of the following forms:

$$F_0 = u, \qquad F_0 = v_n^{n_1}, \qquad F_0 = u_p v_n^{n_p} u_{p-1} v_n^{n_{p-1}} \dots u_1 v_n^{n_1}, \tag{8.7}$$

where $u, u_1, \dots, u_p$ are words that do not contain v_n and v_n^{-1} and are different from e_0, while $n_1, \dots, n_p$ are integers. Construct the required functions and the isomorphism φ by induction. Set, for example, $v_1(x) = xe^x$. Then φ, defined on G_1, maps e only into e_0. Assume that for some $n > 1$ we have already defined functions $v_1(x), \dots, v_{n-1}(x)$ in G and hence φ on G_{n-1} so that φ is an isomorphism on G_{n-1}. First, show that for each word $F \in G_n \backslash G_{n-1}$ there exists a function $v(x, F) = v(x)$ of the form (8.4), depending on F, so that together with already-defined functions $v_1(x), \dots, v_{n-1}(x)$, it satisfies $\varphi(F) \neq e$. It suffices (in view of (8.7)) to consider words F like

$$F = u_p v_n^{n_p} \dots u_2 v_n^{n_2} u_1 v_n^{n_1}, \tag{8.8}$$

where $u_i \in G_{n-1} \backslash e_0$, $i = 1, \dots, p$; $n_1, \dots, n_p$ are non-zero integers.

To shorten the arguments, we can assume that $n_1 > 0$. (If all the exponents $n_k < 0$, then in the word F^{-1} all exponents were positive, and hence $\varphi(F^{-1}) \neq e$ implies $\varphi(F) \neq e$.) We can move this positive exponent in the word F to the extreme right position by circular permutations. Now take an arbitrary (8.4) function $w(x)$ for which the measure $\mu \in \mathcal{P}$. Consider a set of functions (8.4):

$$\Omega = \Omega \begin{pmatrix} x_0 & w(x_0) & \dots & w^{n_1-2}(x_0) \\ w(x_0) & w^2(x_0) & \dots & w^{n_1-1}(x_0) \end{pmatrix}. \tag{8.9}$$

It is non-empty (it contains w). Therefore, the set of values taken at $w^{n_1-1}(x_0)$ by functions v from that set is a non-degenerate interval. Take a value ξ from that interval satisfying the following requirements: ξ is different from the numbers $x_0, w(x_0), \dots, w^{n_1-1}(x_0)$; $\xi \notin u_1^{-1}(x_0, \dots, w^{n_1-1}(x_0))$; $u_1(\xi) \neq \xi$. (The latter can be achieved, since the set of roots of an analytic function $u_1(x)$ is discrete.) Let $w_0(x)$ be one of the functions in Ω for which the value at $w^{n_1-1}(x_0)$ equals ξ. Thus, $w_0^{n_1}(x_0) = w_0(w_0^{n_1-1}(x_0)) = w_0(w^{n_1-1}(x_0)) = \xi$. Consider a non-empty set of functions

$$\Omega_0 = \Omega \begin{pmatrix} x_0 & w_0(x_0) & \dots & w_0^{n_1-2}(x_0) & w_0^{n_1-1}(x_0) \\ w_0(x_0) & w_0^2(x_0) & \dots & w_0^{n_1-1}(x_0) & w_0^{n_1}(x_0) \end{pmatrix} \tag{8.10}$$

(all columns in Ω_0, except for the last "extra" one, coincide with those in Ω) and a point

$$x_1 = u_1 w_0^{n_1}(x_0). \tag{8.11}$$

According to our construction, we have a non-empty set Ω_0 (8.10) and a point x_1 that differs from all the values used in Ω_0. Now apply induction. Assume that we have already constructed a function $w_i(x)$ of the form (8.4) with $\mu \in \mathcal{P}$, a set

$$\Omega_i = \Omega(M_0, \dots, M_i), \tag{8.12}$$

where M_k is an abbreviation for

$$M_k = \begin{pmatrix} x_k & \dots & w_i^{n_k-1}(x_k) \\ w_i(x_k) & \dots & w_i^{n_k}(x_k) \end{pmatrix}, \tag{8.13}$$

for $0 \le k \le i$ when $n_k > 0$ and

$$M_k = \begin{pmatrix} w_i^{-1}(x_k) & \dots & w_i^{n_k}(x_k) \\ x_k & \dots & w_i^{n_k+1}(x_k) \end{pmatrix}$$

for $0 \le k \le i$ when $n_k < 0$, while

$$x_k = u_k w_i^{n_k} u_{k-1} w_i^{n_{k-1}} \dots u_1 w_i^{n_1}(x_0) \tag{8.14}$$

and in the matrices Ω_i all entries are different, so, in particular, all the numbers $x_0, x_1, \dots, x_i$ are distinct. Also, if we set

$$x_{i+1} = u_{i+1} w_i^{n_i}(x_i), \tag{8.15}$$

then the number x_{i+1} differs from all the entries of the matrix Ω_i. Now let us describe the inductive step.

Consider case (a): $n_{i+1} > 0$. Take a function ${}_1w_i(x)$ from Ω_i for which the value ${}_1w_i(x_{i+1})$ differs from all the entries in Ω_i. Set

$$\Omega_i^1 = \Omega\left(M_0, \dots, M_i, \begin{matrix} x_{i+1} \\ {}_1w_i(x_{i+1}) \end{matrix}\right). \tag{8.16}$$

Take a function ${}_2w_i(x) \in \Omega_i^1$ for which the value at ${}_1w_{i+1}(x_{i+1})$ differs from all the entries in Ω_i^1. This generates a new sets of functions:

$$\Omega_i^2 = \Omega\left(M_0, \dots, M_i, \begin{matrix} x_{i+1} & {}_2w_i(x_{i+1}) \\ {}_2w_{i+1}(x_{i+1}) & {}_2w_i(x_{i+1}) \end{matrix}\right). \tag{8.17}$$

Similarly, we construct a function ${}_3w_i(x) \in \Omega_i^2$ and a non-empty set

$$\Omega_i^3 = \Omega\left(M_0, \dots, M_i, \begin{matrix} x_{i+1} & {}_3w_i(x_{i+1}) & {}_3w_i^2(x_{i+1}) \\ {}_3w_i(x_{i+1}) & {}_3w_i^2(x_{i+1}) & {}_3w_i^3(x_{i+1}) \end{matrix}\right) \tag{8.18}$$

so that all entries in Ω_i^3 are distinct, etc. At the end of the chain we form a function ${}_{n_{i+1}}w_i(x) \stackrel{\text{def}}{=} w_{i+1}(x)$ such that all entries in the matrix

$$\begin{gathered} \Omega_{i+1} \stackrel{\text{def}}{=} \Omega_i^{n_{i+1}} = \Omega(M_0, \dots, M_i, M_{i+1}), \\ M_i = \begin{pmatrix} x_{i+1} & \dots & w_{i+1}^{n_{i+1}-1}(x_{i+1}) \\ w_i(x_{i+1}) & \dots & w_{i+1}^{n_{i+1}}(x_{i+1}) \end{pmatrix} \end{gathered} \tag{8.19}$$

are distinct. Moreover, the value $w_{i+1}^{n_{i+1}}(x_{i+1}) = t$ is chosen so that

$$t \notin u_{i+2}^{-1}\left(x_0, \dots, w_{i+1}^{n_{i+1}-1}(x_{i+1})\right),$$

$u_{i+2}(t) = t$. In particular, if we set

$$x_{i+2} = u_{i+2} w_{i+1}^{n_{i+1}}(x_{i+1}), \tag{8.20}$$

this implies that x_{i+2} differs from all the numbers $x_0, \dots, w_{i+1}^{n_{i+1}}(x_{i+1})$, and the induction step from i to $i+1$ is complete.

Consider case (b): $n_{i+1} < 0$. Take a point ξ close to the value $w_i^{-1}(x_{i+1})$ so that x_{i+1} is among the values assumed by functions in Ω_i at that point. According to Lemma 8.2, such a point exists and can be chosen so that it differs from all the numbers used to introduce Ω_i. Let the function ${}_1w_i(x) \in \Omega_i$ be such that ${}_1w_i(\xi) = x_{i+1}$, i.e., $\xi = {}_1w_i^{-1}(x_{i+1})$. We have a set of functions

$$\Omega_i^1 = \Omega\left(M_0, \dots, M_i, \begin{matrix} {}_1w_i^{-1}(x_{i+1}) \\ x_{i+1} \end{matrix}\right) \tag{8.21}$$

and all the numbers in Ω_i^1 are distinct. Now take a point ξ different from all the numbers in Ω_i^1 and close to $w_i^{-2}(x_{i+1})$ so that ${}_1w_i^{-1}(x_{i+1})$ was among the values assumed by the functions Ω_i^1 at ξ. Let ${}_2w_i(x)$ be a function in Ω_i^1 and satisfying ${}_2w_i(\xi) = {}_1w_i^{-1}(x_{i+1})$. Then $\xi = {}_2w_i^{-2}(x_{i+1})$, and we arrive at the set

$$\Omega_i^2 = \Omega\left(M_o, \dots, M_i, \begin{matrix} {}_2w_i^{-1}(x_{i+1}) & {}_2w_i^{-2}(x_{i+1}) \\ x_{i+1} & {}_2w_i^{-1}(x_{i+1}) \end{matrix}\right) \tag{8.22}$$

in which all the numbers are distinct. Continuing this process, we arrive at the function ${}_{|n_{i+1}|}w_i(x) \overset{\text{def}}{=} w_{i+1}(x)$ and the non-empty set

$$\begin{gathered} \Omega_{i+1} \overset{\text{def}}{=} \Omega_i^{|n_{i+1}|} = \Omega_0(M_0, \dots, M_i, M_{i+1}), \\ M_{i+1} = \begin{pmatrix} w_{i+1}^{-1}(x_{i+1}) & w_{i+1}^{-2}(x_{i+1}) & \dots & w_{i+1}^{n_{i+1}}(x_{i+1}) \\ x_{i+1} & w_{i+1}^{-1}(x_{i+1}) & \dots & w_{i+1}^{n_{i+1}+1}(x_{i+1}) \end{pmatrix}, \end{gathered} \tag{8.23}$$

where all the entries are distinct. Moreover, the number $t = w_{i+1}^{n_{i+1}}(x_{i+1})$ can be chosen so that it does not belong to the set $u_{i+2}^{-1}\left(x_0, \dots, w^{n_{i+1}+1}(x_{i+1})\right)$ and $u_{i+2}(t) \neq t$. This, in particular, implies that the number

$$x_{i+2} \overset{\text{def}}{=} u_{i+2} w_{i+1}^{n_{i+1}}(x_{i+1}) \tag{8.24}$$

is different from $x_0, \dots, w_{i+1}^{n_{i+1}}(x_{i+1})$, and the induction step of going from i to $i+1$ has been completed for this case also.

As a result of the above process we obtain a function $w_p(x)$ such that

$$u^p w_p^{n_p} \dots u_1 w_p^{n_1}(x_o) \neq x_0,$$

and so for the word $F = u_p v^{n_p} \dots u_1 v^{n_1}$ we can choose a function $v(x, F) \overset{\text{def}}{=} w_p(x)$ of the form (8.4) with measure $\mu \in \mathcal{P}$ such that on replacing v by $v(x, F)$ we obtain $F(x) \neq x$. Now, enumerate all words in $G_n \backslash G_{n_1} : F_1, F_2, \dots, F_m, \dots$. Take $x_0 > 0$ and construct as above the function $v(x, F_1)$ with $\mu \in \mathcal{P}$. Denote the matrix obtained as before from the values of $v(x, F_1)$ by M_1, and let $\Omega(M_1)$ be a (non-empty) class of functions defined by this matrix. A point y_0 for the construction of

$v(x, F_2)$ is chosen so that all the values that appear when we apply the construction process above lie to the right of those in M_1. This can be achieved as follows. Let

$$F_2 = u_p v_p^{n_p} \dots u_1 r_1^{n_1}$$

and let ρ_0 be the largest entry in the matrix M_1. Among all functions (8.4) the function $v_0(x) = e^{bx}$ has the fastest growth. Hence, if $\rho > \rho_0$, whereas

$$y_0 > \max\left(v_0^{|n_1|}(\rho), u_1 v_0^{|n_1|}(\rho), \dots, v_0^{|n_\rho|}(\rho), u_1 v_0^{|n_1|} \dots u_\rho v_0^{|n_\rho|}(\rho)\right), \tag{8.25}$$

then, starting the process for $v(x, F_2)$ with y_0, we shall only deal with points that lie to the right of ρ. Therefore, we can only deal with functions in $\Omega(M_1)$, and the process will produce the function $v(x, F_2)$ that belongs to $\Omega(M_2)$, where $M_2 \supset M_1$. So, the values of $v(x, F_2)$ at points related to M_1 are the same as those of $v(x, F_1)$. Thus, replacing v_1 in the word F_1 by $v(x, F_2)$, we obtain $F_1(x) \neq x$, in the same way as we did while replacing v_n in that word by $v(x, F_1)$. Continuing the process, we obtain a sequence of functions $\{v(x, F_m)\}$ of the form (8.4) with measures μ_m in $\mathcal{P}$. Also, the functions $v(x, F_m)$ do not lead to $e(x) \equiv x$ when we substitute them into the words $F_1, F_2, \dots, F_m$. Selecting out of measures $\{\mu_m\}$ a sequence that converges weak(*) to the measure μ, we obtain from (8.4) a function $v(x)$. (We cannot now claim that $\mu \in \mathcal{P}$, but this is not needed.) During the construction of $v(x, F_m)$ we obtained the sets $\Omega(M_m)$ for which the lengths of the matrices M_m increase together with m: $M_1 \subset M_2 \subset \dots \subset M_m \subset \cdots$. The function $v(x)$ has in the lowest row of the matrix M_m the same values as $v(x, F_m)$ for $m = 1, \dots,$ and hence when it is substituted into F_m it yields $F_m(x) \not\equiv x$. Thus, the function $v(x) \overset{\text{def}}{=} v_n(x)$ simultaneously "services" all words in $G_n \backslash G_{n-1}$.

Note that in the construction of the functions $v_n(x)$ the segment $[a, b] \subset [0, +\infty)$ was arbitrary. Take a sequence of segments $\{[a_n, b_n]\}$, where $0 < a_{n+1} < b_{n+1} < a_n < b_n \leq 1$ $(b_1 = 1)$. The functions $v_n(x)$ constructed above can be chosen to have the form

$$v_n(x) = x \int_{a_n}^{b_n} e^{xt} d\mu_n(t), \qquad \int_{a_n}^{b_n} d\mu = 1. \tag{8.26}$$

(The value $b_1 = 1$ is chosen only because we have started the construction with $v_1(x) = xe^x$, which corresponds to the unit mass at point 1.) Then

$$\begin{aligned} v_n^{(k)}(x) =& k \int_{a_n}^{b_n} e^{xt} t^{k-1} d\mu_n + x \int_{a_n}^{b_n} e^{xt} t^k d\mu_n < k b_n^{k-1} e^{b_n x} + x b_n^k e^{b_n x} \\ <& k a_{n-1}^{k-1} e^{a_{n-1} x} + x a_{n-1}^k e^{a_{n-1} x} < v_{n-1}^{(k)}(x). \end{aligned} \tag{8.27}$$

The inequality (8.27) completes the proof of the theorem.

THEOREM 8.3. *There exist compact sets $X, X_1, \dots, X_n, \dots$ and surjective mappings $\varphi_i : X \to X_i$, $i = 1, \dots, n, \dots,$ satisfying the following properties:*

1. *For any three distinct indices i_1, i_2, i_3 we have for an arbitrary function $g(x) \in B(X)$ the following representation:*

$$\begin{aligned} g(x) = g_1 \circ \varphi_{i_1}(x) + g_2 \circ \varphi_{i_2}(x) + g_3 \circ \varphi_{i_3}(x), \\ g_k \in B(X_k), \qquad k = 1, 2, 3. \end{aligned} \tag{8.28}$$

2. *There exists $f(x) \in C(X)$ that cannot be represented in the form*

$$f(x) = \sum_{1}^{\infty} g_i \circ \varphi_i(x), \qquad g_i \in C(X_i), \quad \sum_{1}^{\infty} \|g_i\| < +\infty. \tag{8.29}$$

3. *For any two indices $i_1 \neq i_2$, the subspace*

$$g_1 \circ \varphi_{i_1}(x) + g_2 \circ \varphi_{i_2}(x), \qquad g_1 \in C\left(X_{i_1}\right), g_2 \in C\left(X_{i_2}\right) \tag{8.30}$$

is dense in $C(X)$.

PROOF. *Construction.* Let $\{v_n(x)\}$ be a sequence of functions constructed in the proof of Theorem 8.1. There are countably many words in $G_0 \backslash e_0$, and for each word F, in view of analyticity of the function $F(x)$ (i.e., $(\varphi f)(x)$, where φ is the isomorphism constructed in Theorem 8.1), there are at most countably many roots of the equation $F(x) = x$. Let S_0 be the set of all the roots of all the equations for all words. Since S_0 is a countable set, it can be covered by open intervals with an arbitrarily small total length. Let S be the union of such intervals covering S_0. It is easy to choose it so that

$$\operatorname{meas}\left[S \cap [0, h]\right] = o(h), \qquad h \to 0^+. \tag{8.31}$$

Let $X = [-1, 1] \backslash S$. Since S is open, X is compact. Define the functions $\varphi_n(x)$ on X, $n = 1, 2, \ldots$, by setting

$$\varphi_n(x) = x \quad \text{when } x \geq 0, \qquad \varphi_n(x) = v_n(|x|) \quad \text{when } -1 \leq x < 0. \tag{8.32}$$

Let $X_n = \varphi_n(X)$. Since $v_n(1) > 1$ (in view of (8.2) and (8.6)), $X_n = [0, v_n(1)]$. Thus,

$$X = [-1, 1] \backslash S, \qquad X_n = \varphi_n(X), \qquad \varphi_n(x) = \begin{cases} x, & x \geq 0, \\ v_n\left(|x|\right), & x < 0. \end{cases} \tag{8.33}$$

Note the very simple structure of the compact sets X and $\{X_n\}$. In particular, the X_n are simply segments.

PROOF OF PROPERTY 1. Call the points x and x' in X equivalent with respect to R_i, $i = 1, 2, \ldots$, if $\varphi_i(x) = \varphi_i\left(x'\right)$. We shall write it as xR_ix'. If $x \neq x'$ and xR_ix', then x and x' have opposite signs and $x' = v_i\left(|x|\right)$ when $x < 0$, whereas $|x'| = v_i^{-1}(x)$ when $x > 0$. Clearly, R_i is indeed an equivalence relation. Consider a sequence of points $x_1, x_2, \ldots, x_n$ of a set X and certain relations $R_{i_1}, R_{i_2}, \ldots, R_{i_{n-1}}$. Let

$$x_1 R_{i_1} x_2 R_{i_2} x_3 \ldots x_{n-1} R_{i_{n-1}} x_n. \tag{8.34}$$

(8.34) means that $x_1 R_{i_1} x_2$, $x_2 R_{i_2} x_3$, etc. Assume that $x_1 \neq x_2, x_2 \neq x_3, \ldots, x_{n-1} \neq x_n$ and, also, $i_1 \neq i_2, i_2 \neq i_3, \ldots, i_{n-2} \neq i_{n-1}$ (i.e., the consecutive relations R_i in (8.34) are all distinct). Under these assumptions call (8.34) a lightning bolt joining x_1 and x_n (concerning the origin of this terminology, cf. Chapter 2, where this notion plays a crucial role). Every point $x \in X$ is considered to be a lightning bolt joining x with x. Let us show that two points x and x' can be joined by not more than one lightning bolt. Let x_1 and x_n be joined by the lightning bolt (8.34). Relate to it the word F in the free group G_0:

$$F = v_{i_{n-1}}^{k_{n-1}} \ldots v_{i_2}^{k_2} v_{i_1}^{k_1},$$

where the exponents k_j alternate between 1 and -1: $k_1 = 1$ if $x_1 < 0$ and $k_1 = -1$ if $x_1 > 0$. Under our assumptions F is indeed a word, since these assumptions provide for the lack of "empty" words in F. In case when $x_1 = x_n$, we take $F = e_0$. The word F under the isomorphism φ corresponds to a function $F(x) = (\varphi F)(x)$. Then, $|x_n| = F(|x_1|)$. Suppose there is another lightning bolt that joins x_1 and x_n. It corresponds to another word $H \in G_0$ which, in turn, corresponds to a function $H(x) \in G$. Then $F(|x_1|) = H(|x_1|)$. If $x_1 > 0$, it follows that $H^{-1} \circ F(x_1) = x_1$. But by construction X does not contain roots of any equation $\Phi(x) = x$ when $\Phi \in G_0 \backslash e_0$. Hence, $H^{-1}F = e_0$ and, therefore, the words H and F, and the corresponding lightning bolts, are the same. If $x_1 < 0$, then $x_2 > 0$ and $x_2 = v_{i_1}(|x_1|)$, $|x_1| = v_{i_1}^{-1}(x_2)$. We obtain $F \circ v_{i_1}^{-1}(x_2) = H \circ v_{i_1}^{-1}(x_2)$ and, as in the previous case, $Fv_{i_1}^{-1} = Hv_{i_1}^{-1}$, $F = H$. (In addition, if we had $x_1 = x_n$, then taking $F = e_0$ we would have obtained $H = e_0$ as well.)

Now select among the mappings $\{\varphi_n\}$ three with different indices. To fix the ideas, let them be $\varphi_1, \varphi_2, \varphi_3$. In addition to the equivalence relations R_1, R_2, R_3 defined above, introduce on X one more relation R by setting xRx' if x and x' can be joined by a lightning bolt that only contains relations R_1, R_2, R_3. Without any difficulties one can show that R is an equivalence relation. Let $h_i = g_i \circ \varphi_i$, $g_i \in B(X_i)$. Clearly, $h_i(x) \in B(X)$, and it takes constant values on each equivalence class in the quotient set X/R_i. Conversely, if the $h_i(x)$ are in $B(X)$ and assume constant values on each equivalence class in X/R_i, then obviously one can find $g_i \in B(X_i)$ such that $h_i = g_i \circ \varphi_i$. Hence, for $f(x) \in B(X)$ we must find bounded functions $H_i(x)$, $i = 1, 2, 3$, such that

$$f = h_1 + h_2 + h_3, \qquad h_i(x) = h_i(x') \quad \text{if } xR_ix'.$$

Now consider the equivalence classes X/R. Each such class is a saturation class for X/R_i. The latter means that if E is some class in X/R, $x \in E$ and $X'R_ix$, then $x' \in E$. Therefore, functions h_i can be defined on each class in X/R independently of their definition on other classes in X/R as long as h_i remains bounded on all of X.

Let E be an equivalence class in X/R. Choose a point $x_1 \in E$. For any $x \in E$ call the number of points in the lightning bolt joining x_1 to x, the rank of x. Set $h_1(x_1) = f(x_1)$, $h_2(x_1) = h_3(x_1) = 0$. Suppose that the values of h_i are defined for all points of rank less than n ($n > 1$) so that at those points $|h_i| \le \|f\|$, if xR_ix' then $h_i(x) = h_i(x')$, and, finally, $f = h_1 + h_2 + h_3$. Let x_n be a point of rank n. It is joined to x_1 by the lightning bolt (8.34), where now all $i_k = 1, 2, 3$. The point x_{n-1} had rank $n-1$ and, in view of the uniqueness of a lightning bolt joining x_1 and x_n, this is the only point of rank $n-1$ that is equivalent to x_n with respect to one of the relations R_i, $i = 1, 2, 3$. (Otherwise, there would be two lightning bolts joining x_1 and x_n, or x_1 could be joined to x_n by a lightning bolt with less than n points.) Let $x_{n-1}R_ix_n$, and let $j \ne k$ be the remaining two numbers among $1, 2, 3$ different from i. Set $h_i(x_n) = h_i(x_{n-1})$, $h_j(x_n) = -h_i(x_n)$, $h_k(x_n) = f(x_n)$. Continuing this process inductively (with respect to the rank), we properly define h_1, h_2, h_3 on the whole class $E \in X/R$. Combining the results over all such classes, we obtain on X

$$f = h_1 + h_2 + h_3, \quad |h_i(x)| \le \|f\|,$$
$$h_i(x) = h_i(x') \quad \text{whenever } xR_ix', \qquad i = 1, 2, 3.$$

Using h_i, we find appropriate $g_i \in B(X_i)$, and hence f has the form (8.28).

PROOF OF PROPERTY 2. If for each $f(x) \in C(X)$ (8.29) is satisfied, then according to Theorem 2.1 there exists a number λ $(0 < \lambda \le 1)$ such that for each $\mu \in C(X)^\star$,

$$\sup_n \|\varphi_n \circ \mu\| \ge \lambda \|\mu\|. \tag{8.35}$$

Denote by m the Lebesgue measure on the real line, and, taking h, $0 < h < 1$, define a measure μ on Borel subsets $E \subset X$ by setting

$$\begin{aligned} &\mu(E) = \frac{m(E)}{h}, \qquad E \subset [0,h], \\ &\mu(E) = -\frac{m(E)}{h}, \qquad E \subset [-h,0], \\ &\mu(E) = 0, \qquad E \subset [-1,-h) \cup (h,1]. \end{aligned} \tag{8.36}$$

Then,

$$\|\mu\| = \frac{1}{h}\left[2h - m\left(S \cap [0,h]\right)\right] = 2 - o(1). \tag{8.37}$$

Take an arbitrary $\varepsilon > 0$ and choose h so small that

$$v_1'(x) < 1 + \varepsilon, \qquad 0 < x < h. \tag{8.38}$$

For all n, (8.6) yields the following inequalities:

$$\begin{aligned} &1 < v_n'(x) < v_1'(x), \qquad x < v_n(x) < v_1(x), \qquad x > v_n^{-1}(x), \\ &\left(v_n^{-1}\right)'(x) = \frac{1}{v_n'\left(v_n^{-1}(x)\right)} > \frac{1}{v_n'(x)} > \frac{1}{v_1'(x)} > \frac{1}{1+\varepsilon} > 1 - \varepsilon. \end{aligned} \tag{8.39}$$

Now estimate $\|\varphi_n \circ \mu\|$. According to Lemma 3.2 we have:

$$\|\varphi_n \circ \mu\| = \|\mu\| - 2\Theta_n, \tag{8.40}$$

where Θ_n is defined as in that lemma. According to (8.36), we have for A (in the notation of Lemma 3.2) $X \cap [0,h]$, $B = [-h,0]$. In view of the definition of φ_n we obtain

$$C = \varphi_n(A) \cap \varphi_n(B) = A \cap [0, v_n(h)] = A = [0,h] \backslash S. \tag{8.41}$$

Let $P \subset C$. Then (using the notation in Lemma 3.2, and (8.36), (8.39)),

$$\begin{aligned} \Theta_n(P) &= \min\left\{\mu(P'), |\mu(P'')|\right\} = \min\left\{\mu(P), \left|\mu\left(v_n^{-1}(P)\right)\right|\right\}; \\ \mu(P) &= \frac{1}{h} m(P) \left|\mu\left(v_n^{-1}(P)\right)\right| = \frac{1}{h}\int_P \left(v_n^{-1}\right)' dm > \frac{1}{h}(1-\varepsilon) m(P). \end{aligned}$$

Thus,

$$\Theta_n(P) > \frac{1-\varepsilon}{h} m(P). \tag{8.42}$$

For an arbitrary partition of the set C into sets P_i, $i = 1, \dots, k$, we obtain from (8.4) and (8.42)

$$\sum_1^k \Theta_n(P_i) > \frac{1-\varepsilon}{h} \sum_1^k m(P_i) = \frac{1-\varepsilon}{h}[h - o(h)] = (1-\varepsilon)(1-o(1)).$$

Therefore, for Θ_n in (8.40) we obtain the estimate

$$\Theta_n = \inf \sum_1^k \Theta_n(P_i) > 1 - \varepsilon - o(1) > \frac{1}{2}\|\mu\| - \varepsilon. \tag{8.43}$$

(8.35) is equivalent (Lemma 4.3) to the inequality

$$\sup_n 2\Theta_n \le (1-\lambda)\|\mu\|.$$

But (8.43) yields that the latter inequality does not hold for any λ $(0 < \lambda \le 1)$ if we take h sufficiently small.

PROOF OF PROPERTY 3. Take, for example, $i = 1, 2$. We must show that the subspace D

$$D = \{g_1 \circ \varphi_1(x) + g_2 \circ \varphi_2(x)\}, \qquad g_1 \in C(X_1), \quad g_2 \in C(X_2) \tag{8.44}$$

is everywhere dense in $C(X)$. Let $\mu \in D^\perp$. Using Fubini's theorem, we right away obtain that $\mu \in D^\perp$ if and only if $\nu_i = \varphi_i \circ \mu = 0$, $i = 1, 2$. We need to show that in our situation from $\nu_1 \equiv 0$ and $\nu_2 \equiv 0$ it follows that $\mu \equiv 0$. Assume the opposite. Consider intervals $(t, 1]$, $t \ge -1$, for which the variation of the measure μ on $(t, 1] \cap X$ equals zero. (The empty interval $(1, 1]$ is also allowed.) Let $(T, 1]$ be the union of all such intervals. Assume that $T > 0$. Since for $x > 0$ (cf. (8.39)) $v_2(x) < v_1(x)$, it follows that $v_1^{-1}(T) < v_2^{-1}(T)$ and $v_2 \circ v_1^{-1}(T) < T$. Consider a closed set P in $\left[-1, -v_1^{-1}(T)\right]$. Then the set $Q = \varphi_1(P) \cap X$ is also closed. By definition of φ_1 we have $Q = v_1(-P) \cap X$ ($-P$ is the set symmetric to P with respect to the origin) and $Q \subset [T, 1) \cap X$. Moreover, $\varphi_1^{-1}(Q) = Q \cup P$. Since we assumed $\nu_1 \equiv 0$, we obtain

$$\nu_1(Q) \stackrel{\text{def}}{=} \mu\left(\varphi_1^{-1}(Q)\right) = \mu(Q) + \mu(P) = 0.$$

However, $\mu(Q) = 0$ also, since $Q \subset [T, 1) \cap X$. Hence, $\mu(P) = 0$. Now let M be a closed set inside $\left(v_1 \circ v_1^{-1}(T), 1\right] \cap X$ and $N = -v_2^{-1}(M)$. Then N is also closed, and $N \subset \left[-1, -v_1^{-1}(T)\right]$. Therefore, in view of what we have shown above, $\mu(N) = 0$. Furthermore, $\varphi_2^{-1}(M) = M \cup N$. Since $\nu_2 \equiv 0$ by our assumption, whereas

$$\nu_2(M) \stackrel{\text{def}}{=} \mu\left(\varphi_2^{-1}(M)\right) = \mu(M) + \mu(N) = 0,$$

then $\mu(M) = 0$ as well. Thus, on an arbitrary closed set $M \subset \left(v_2 \circ v_1^{-1}(T), 1\right]$ the measure μ vanishes, $\mu(M) = 0$.

In view of the regularity of μ this implies that the total variation of μ on $(v_2 \circ v_1^{-1}(T), 1]$ vanishes. But we have already noted that $v_2 \circ v_1^{-1}(T) < T$ and, hence, the variation of μ vanishes on a semi-interval that is larger than $(T, 1]$. This contradicts the definition of T. The contradiction appeared because we assumed that $T > 0$. Hence $T \le 0$. But then, recalling again that $\nu_1 \equiv 0$ and $\nu_2 \equiv 0$, we conclude that $\mu \equiv 0$, and the proof of the theorem is now complete.

REMARK. We have already noted that in [**136**] the examples are constructed with $BD = B(X)$, but $D \neq C(X)$. However, in those examples $\overline{D} \neq C(X)$. So, both possibilities may occur:

1. $BD = B(X)$, $D \neq C(X)$, $\overline{D} = C(X)$;
2. $BD = B(X)$, $\overline{D} \neq C(X)$.

§9. Measure of compact sets on which all continuous functions are representable by sums of superpositions

Let X, X_i, $i = 1, \dots, N$, be compact sets, $\varphi_i : X \to X_i$ continuous mappings. Consider the products

$$Y = X_1 \times \cdots \times X_N, \quad Y_k = X_1 \times \cdots \times X_k, \quad \tilde{Y}_k = X_{k+1} \times \cdots \times X_N, \qquad k < N. \tag{9.1}$$

Assume that the family of mappings $F = (\varphi_1, \dots, \varphi_N)$ separates points in X. Then, the mapping

$$F : X \to Y, \qquad x \in X \to (\varphi_1(x), \dots, \varphi_N(x)) \tag{9.2}$$

is a homeomorphism. Identifying X with its image $F(X)$ in Y, we can assume $X \subset Y$ and the φ_i are natural projections from Y onto X_i.

Also, consider the mappings

$$\Phi : X \to Y_k \quad \text{and} \quad \tilde{\Phi} : X \to \tilde{Y}_k \tag{9.3}$$

that are the restrictions to X of the natural projections of Y onto Y_k and $\tilde{Y}_k$, respectively.

THEOREM 9 ([**131**]). *Let all $f \in C(X)$ be representable in the form*

$$f(x) = g \circ \Phi(x) + \tilde{g} \circ \tilde{\Phi}(x), \qquad g \in C(Y_k), \quad \tilde{g} \in C\left(\tilde{Y}_k\right). \tag{9.4}$$

If λ and μ are two arbitrary positive finite measures on Y_k and $\tilde{Y}_k$ respectively that vanish on one-point sets and $\nu = \lambda \times \mu$, then

$$\nu(X) = 0. \tag{9.5}$$

In particular, (9.5) *holds if for all $f \in C(X)$ the representation* (3.18) *holds.*

PROOF. Assume that $\nu(X) > 0$. By Fubini's theorem we have:

$$\nu(X) = \int_{Y_k} \mu\left[\tilde{\Phi}\left(\Phi^{-1}(t)\right)\right] d\lambda_t.$$

Therefore, there exists an uncountable set of points $t \in Y_k$ for which

$$\mu\left[\tilde{\Phi}\left(\Phi^{-1}(t)\right)\right] > 0.$$

Hence, there exist a positive number η and an infinite sequence of distinct points $\{t_i\}$, $i = 1, \dots, n, \dots$, in Y_k for which

$$\mu\left(E_i\right) > \eta, \qquad E_i = \tilde{\Phi}\left(\Phi^{-1}\left(t_i\right)\right). \tag{9.6}$$

Among them there exist at least two points t_i and t_j such that $E_i \cap E_j$ contains at least two points. Indeed, assuming that every pair E_i and E_j intersects over at most one point, we form the set $E = \bigcup E_i$. Since the measure μ has no atoms, it follows that

$$\mu(E) = \sum_i \mu(E_i)$$

and according to (9.6), $\mu(E) = \infty$, contradicting the finiteness of the measure μ.

Thus, let $E_i \cap E_j$ contain two points α and β from $\tilde{Y}_k$. Consider four points in the set X

$$x_1 = (t_i, \alpha), \qquad x_2 = (t_i, \beta), \qquad x_3 = (t_j, \alpha), \qquad x_4 = (t_j, \beta)$$

and the measure

$$\delta = \delta_{x_1} - \delta_{x_2} - \delta_{x_3} + \delta_{x_4}, \tag{9.7}$$

where δ_x denotes the delta-measure at the point x. A straightforward calculation yields that the measure (9.7) annihilates all functions (9.4), and hence $C(X)$ cannot consist only of such functions (e.g., a function $f(x)$ such that $f(x_1) - f(x_2) + f(x_3) - f(x_4) \neq 0$ cannot be represented in the form (9.4)).

CHAPTER 2

Approximation of Functions of Two Variables by Sums $\varphi(x) + \psi(y)$

§1. Raising the questions. Lightning bolts

Asking the questions. Let D be the subspace of linear superpositions in $C(X)$. In Chapter 1 we studied various approaches to the following question:

1. When is $D = C(X)$?

In the case of bounded functions the question was posed on coincidence of the subspace BD with $B(X)$. In the case when $D \neq C(X)$ we are now going to study the problem of best approximation of functions $f \in C(X)$ by elements of the subspace D (or of functions $f \in B(X)$ by the elements of the subspace BD). This gives rise to the following questions:

2. When is D everywhere dense in $C(X)$?
3. When is D closed in $C(X)$?
4. When is D proximinal in $C(X)$?

Recall that a set W in a metric space V is called proximinal if for each element $v \in V$ there exists an element $w_0 \in W$ closest to v among all elements of W.

The above questions are also natural in the context of the subspace BD in $B(X)$. Concentrate on subspaces of sums of superpositions—the simplest in structure ((2.1) or (2.6) of Chapter 1). Thus, we are considering the subspaces D and BD that consist of functions

$$g_1 \circ \Phi_1(x) + \cdots + g_N \circ \Phi_N(x), \tag{1.1}$$

where $X, X_1, \ldots, X_N$ are compact sets, $\Phi_i : X \to X_i$, $i = 1, \ldots, N$, are continuous mappings and $g_i \in C(X_i)$ are arbitrary (in case of the subspace D), or $X, X_1, \ldots, X_N$, $\Phi_i : X \to X_i$ are arbitrary sets and mappings while $g_i \in B(X_i)$ (the subspace BD).

Reformulation in terms of function algebras. Consider the subspace D of functions (1.1). Functions

$$g_i \circ \Phi_i(x), \qquad g_i \in C(X_i), \tag{1.2}$$

form a closed subalgebra A_i of the algebra $C(X)$, and A_i contains the constants. Conversely, let A_i, $i = 1, \ldots, N$, be closed subalgebras of $C(X)$ (X is still compact) containing the constants. Define the equivalence relation R_i, $i = 1, \ldots, N$, for points in X by setting

$$xR_ix' \quad \text{if } f(x) = f(x') \quad \text{for all } f \in A_i. \tag{1.3}$$

Then $X_i = X/R_i$, the quotient space of X with respect to the relation R_i equipped with the quotient space topology, is compact and the natural projection $\Phi_i : X \to$

X_i is continuous (see [**1**], [**117**]). The space X_i is not merely compact but is also Hausdorff; hence (being compact) it is a normal space. This is seen from the following lemma.

LEMMA 1.1. *Let A be a family of functions continuous on a compact space X, and r an equivalence relation defined by A:*

$$xrx' \Leftrightarrow f(x) = f(x') \qquad \forall f \in A. \tag{1.4}$$

The saturation $r(F)$ of any closed set $F \subset X$ ($r(F) = \bigcup_{x\in F} r(x)$; $r(x)$ is the equivalence class of a point x) is closed, and the canonical projection $\Phi : X \to X/r$ is a closed mapping ($r(F) = \Phi^{-1}(\Phi(F))$).

PROOF. Let x_0 be a limit point of $r(F)$. There exists a net $\{x_\alpha\} \subset r(F)$ converging to x ($\{\alpha\}$ is a directed set of indices). For each x_α there exists $y_\alpha \in F$ with $y_\alpha r x_\alpha$. Since F is compact, we can select from the net $\{y_\alpha\}$ a subnet $\{y_\beta\} \to y_0 \in F$. The corresponding subnet $\{x_\beta\} \to X_0$, and for any $f \in A$ we have $f(x_0) = \lim f(x_\beta) = \lim f(y_\beta) = f(y_0)$, i.e., $x_0 r y_0$ and $x_0 \in r(F)$. Closedness of the projection Φ is equivalent to what we have proved (cf. [**117**]).

COROLLARY 1.2. *Under the assumptions of Lemma* 1.1, *X/r is a normal compact space and therefore is a Hausdorff compact space.*

For a proof, cf. [**117**].

In view of the Stone–Weierstrass theorem ([**40**]), which applies here since X/R_i is compact,

$$A_i = \{g_i \circ \Phi_i(x),\ \forall g_i \in C(X_i)\}. \tag{1.5}$$

Hence, the questions raised above can be reformulated as follows. Let A_i, $i = 1, \dots, N$, be closed subalgebras in $C(X)$ containing the constants.

1. When does $A_1 + \dots + A_N$ coincide with $C(X)$?
2. When is $A_1 + \dots + A_N$ dense in $C(X)$?
3. When is $A_1 + \dots + A_N$ closed in $C(X)$?
4. When is $A_1 + \dots + A_N$ proximinal?

Changing the set-up. Suppose (this is natural for a number of situations) that the mapping

$$\Psi : X \to X_1 \times \dots \times X_N, \qquad x \to \Psi(x) = (\Phi_1(x), \dots, \Phi_N(x)), \tag{1.6}$$

is injective. Then Ψ is a homeomorphism between X and $Q = \Psi(X)$, and all the questions can be studied for $Q \subset X_1 \times \dots \times X_N$, treating the Φ_i as natural projections of Q into X_i.

In this chapter we shall study the case $N = 2$. On the one side, there is a meaningful theory for that case. On the other side the case $N > 2$ is very little studied, and the difficulties here are much greater. In relation to question 1, we have already seen in §8 of Chapter 1 that there is an essential difference between the cases $N = 2$ and $N > 2$.

The simplest case. For the problem with $N = 2$ it is worthwhile to change the notation: Q, X, Y are compact sets, points in Q are denoted by p, q, etc. $\pi_1 : Q \to X$, $\pi_2 : Q \to Y$ are continuous mappings. Then,

$$D = D(Q) = \{g_1 \circ \pi_1(p) + g_2 \circ \pi_2(p)\}, \qquad g_1 \in C(X), \quad g_2 \in C(Y). \tag{1.7}$$

Clearly, $D(Q) \subset C(Q)$. In the case when $Q \subset X \times Y$, and π_1 and π_2 are natural projections from $X \times Y$ onto X and Y respectively, for $p = (x, y)$ we set $\pi_1(p) = x$, $\pi_2 = y$; $x \in X$, $y \in Y$. In that case, we can write the functions in $D(Q)$ in a simpler form:

$$D(Q) = \{\varphi(x) + \psi(y),\ \varphi \in C(X),\ \psi \in C(Y)\}. \tag{1.8}$$

We shall call this set-up the simplest (one can assume that $\pi_1(Q) = X$ and $\pi_2(Q) = Y$). The main tendencies for $D(Q)$ are already seen in the special case when $Q \subset \mathbb{R}^2$ and x, y are the usual coordinates in $\mathbb{R}^2$.

Lightning bolts. Let $\pi_1 : Q \to X$ and $\pi_2 : Q \to Y$ be continuous mappings of the compact set Q onto the compact sets X and Y, respectively. Introduce on Q equivalence relations R_1, R_2 by setting pR_iq if $\pi_i(p) = \pi_i(q)$. If we consider the subalgebras $A_i = \{g_i \circ \pi_i(p)\}$, $i = 1, 2$, where g_1, g_2 are arbitrary functions in $C(X)$, $C(Y)$ respectively, then

$$\pi_i(p) = \pi_i(q) \Leftrightarrow f(p) = f(q), \qquad f \in A_i. \tag{1.9}$$

A finite or infinite sequence ℓ of points $[p_1, p_2, \dots, p_n, \dots]$ in Q is called a *lightning bolt* (respectively finite or infinite) if $p_i \neq p_{i+1}$, $i = 1, \dots,$ and either

$$p_1R_1p_2, p_2R_2p_3, p_3R_1p_4, \dots$$

or

$$p_1R_2p_2, p_2R_1p_3, p_3R_2p_4, \dots. \tag{1.10}$$

The points p_i, $i = 1, \dots,$ are called vertices of the lightning bolt. The point p_1 is the beginning of the lightning bolt. If the lightning bolt $\ell = [p_1, \dots, p_n]$ is finite, then p_n is its end and ℓ joins p_1 to p_n. Every point p can be considered as a lightning bolt joining p to p. If a lightning bolt $\ell = [p_1, \dots, p_n]$ is finite, the number n of its vertices is called its length. A point is considered to be a lightning bolt of length one. Sometime we shall specify the relations R_i in the lightning bolt notation, e.g.,

$$p_1R_1p_2R_2p_3R_1p_4R_2p_5R_1p_6. \tag{1.11}$$

In the case when $Q \subset \mathbb{R}^2$ and π_1, π_2 are coordinate projections, each segment p_ip_{i+1} in the lightning bolt $[p_1, p_2, \dots]$ is parallel either to the Ox-axis or Oy-axis, and two consecutive segments $p_{i-1}p_i$ and p_ip_{i+1} are perpendicular. This easily-visualized interpretation is useful in a general situation as well. The notion of a lightning bolt has already appeared in §8 of Chapter 1. We shall see that for $N = 2$ the notion of a lightning bolt turns out to be very useful for answering the questions raised above, concerning the subspaces D and BD.

The notion of a lightning bolt appeared in the works of Arnold [**4**], [**5**] where the thirteenth Hilbert problem was solved. Arnold considered lightning bolts in $\mathbb{R}^3$. Later the notion of a lightning bolt has been used in practically all works dealing with representation and approximation by superpositions, though sometimes it appeared under different names ([**113**], [**99**], [**100**],[**79**], [**80**], [**61**], [**91**], and others).

Circular and closed lightning bolts. A finite lightning bolt $[p_1, \dots, p_n]$ is called closed if $p_1R_1p_2$ and $p_nR_2p_1$, or $p_1R_2p_2$ and $p_nR_1p_1$. A closed lightning bolt must have an even number of vertices: $n = 2m$. A finite lightning bolt $[p_1, \dots, p_n]$ is called circular if $p_n = p_1$ and at least two vertices of the bolt are distinct. If $[p_1, \dots, p_{2m}]$ is a closed lightning bolt, then $[p_1, \dots, p_{2m}, p_1]$ is circular.

If $[p_1,\dots,p_n,p_1]$ is a circular lightning bolt, then $[p_1,\dots,p_n]$ need not be closed, because the equivalences of p_1 with p_2 and p_n with p_1 for n odd turn out to be the same. Yet from a circular lightning bolt one can select a closed one by omitting some vertices. Closed lightning bolts were used (under a different name) in [**38**] even before the general notion of a lightning bolt had evolved.

Irreducible lightning bolts. A (finite) lightning bolt is called irreducible if there does not exist a lightning bolt of a smaller length with the same beginning and end. A lightning bolt $[p_1,p_2]$ is always irreducible ($p_1\neq p_2$). A lightning bolt containing a single vertex p is irreducible. In an irreducible lightning bolt all vertices are distinct.

The relation R. Define yet one more relation R: pRq if there exists a finite lightning bolt joining p and q.

LEMMA 1.3. *R is an equivalence relation. The equivalence class $R(p)$ of a point p is an F_σ-set. If E is a closed subset in Q, then its saturation $R(E)$ is an F_σ-set. The saturation of a Borel set is a Borel set.*

PROOF. Clearly pRp, and if pRq, then qRp. Let us show that $pRq, qRs\Rightarrow pRs$. Let a lightning bolt $[p_1,\dots,p_n]$ join p and q ($p_1=p, p_n=q$) and let a lightning bolt $[q_1,\dots,q_m]$ join q and s ($q_1=q, q_m=s$). If $p_{n-1}R_{i_1}p_n$ and $q_1R_{i_2}q_2$, $i_1\neq i_2$ ($i_1,i_2=1,2$), then the lightning bolt $[p_1,\dots,p_n,q_1\dots,q_m]$ joins $p_1=p$ and $q_m=s$. If, on the other hand, $p_{n-1}R_ip_n$ and $q_1R_iq_2$, the lightning bolt $[p_1,\dots,p_{n-1},q_2,\dots,q_m]$ joins $p_1=p$ and $q_m=s$. Let $E\subset Q$ be closed, and hence, compact. Then the sets

$$E_1=\pi_1^{-1}(\pi_1(E)),E_2=\pi_2^{-1}(\pi_2(E_1)),E_3=\pi_1^{-1}(\pi_1(E_2)),\dots$$

are all compact. It is easy to see that $R(E)=\bigcup_{k=1}^{\infty}E_k$; hence $R(E)$ is F_σ. In particular, if $E=\{p\}$, we obtain that the equivalence class $R(p)$ is a F_σ-set. If E is a Borel set, then $E_1,\dots,E_k,\dots$ are also Borel sets, and so is $R(E)$.

Relation R_3. In addition to the equivalence relations R_1,R_2,R on Q introduced above, we shall need one more. Consider the algebra $A_3=A_1\cap A_2$. It is a closed subalgebra in $C(Q)$ that contains constants. Set

$$pR_3q,\quad \text{if } f(P)=f(q)\quad \text{for all } f\in A_3. \tag{1.12}$$

Obviously, $R(p)\subset R_3(p)$.

The condition of representability by superpositions in terms of lightning bolts. Consider an operator $\tau(Z)$ defined by the formulas (7.1), (7.2) of Chapter 1.

PROPOSITION 1.4. *For any $Q,X,Y,\pi_1:Q\to X,\pi_2:Q\to Y$, the following are equivalent.*

1. *There exists a natural number $n=n(Q)$ such that $\tau^n(Q)=Q$.*
2. *Q does not contain closed lightning bolts, and there exists $M=M(Q)$ such that every irreducible lightning bolt $\ell\subset Q$ has no more than M vertices.*

PROOF. Let us show that 1 $\Rightarrow$ 2. For a closed lightning bolt ℓ, we have $\tau(\ell)=\ell$, hence $\tau^n(\ell)=\ell$ for all n. Therefore, if Q contains a closed lightning bolt, $\tau^n(Q)=Q$ is impossible. Moreover, if there exists an irreducible lightning bolt ℓ in Q with k vertices, then $\tau(\ell)$ is obtained from ℓ by omitting the first and last vertices, and therefore $\tau(\ell)$ is a lightning bolt with $k-2$ vertices. In general, $\tau^n(\ell)$ contains $k-2n$ vertices, and since $\tau^n(Q)=\ell$, $k\leq 2n$.

Let us show that 2 $\Rightarrow$ 1. Since Q contains no closed (hence, circular) lightning bolts, if two points are joined by a lightning bolt it is unique, hence irreducible. Let $\tau^n(Q)\neq\emptyset$ and $p_1\in\tau^n(Q)$. Then there exists $p_2\neq p_1$, $p_2\in\tau^{n-1}(Q)$, $p_1R_1p_2$. Moreover, there exists $p_3\in\tau^{n-2}(Q)$, $p_3\neq p_2$, $p_2R_2p_3$, etc. Finally, there exists $p_n\in\tau(Q)$ and $[p_1,\dots,p_n]$ is a lightning bolt on Q that is not circular and is irreducible. Then $n\leq M$, while for $n>M$ necessarily $\tau^n(Q)=\emptyset$.

We can now reformulate Theorems 7.3 and 7.4 of Chapter 1.

THEOREM 1.5. *Let Q,X,Y be arbitrary sets, and $\pi_1:Q\to X$, $\pi_2:Q\to Y$ be arbitrary mappings. In order that the subspace*

$$BD=\{g_1\circ\pi_1(p)+g_2\circ\pi_2(p)\}\,,\qquad g_1\in B(X),\quad g_2\in B(Y)$$

coincide with $B(Q)$ it is necessary and sufficient that Q contain no closed lightning bolts and the lengths of all irreducible lightning bolts be bounded by the same number.

THEOREM 1.6. *Let Q,X,Y be compact metric spaces, and $\pi_1:Q\to X$, $\pi_2:Q\to Y$ continuous mappings. For a subspace D (1.7) to coincide with $C(Q)$ it is necessary and sufficient that Q contain no closed lightning bolts and the lengths of all irreducible lightning bolts be bounded by the same number.*

This result has been obtained by the author in [**80**] independently of [**133**], and the theorem remains true for arbitrary compact spaces Q,X,Y, not merely compact metric spaces. This will be shown later on in §4.

Now let us raise a natural question complementing question 1. Again, let $X,X_1,\dots,X_N$ be sets, and $\Phi_i:X\to X_i$ be mappings. Under what conditions can an arbitrary function $f(x)$ defined on X be represented in the form

$$f(x)=g_1\circ\Phi_1(x)+\cdots+g_N\circ\Phi_N(x),\tag{1.13}$$

similar to (1.1), but with arbitrary g_i defined on X_i? Thus, no boundedness or continuity assumptions are implemented on $f,g_1,\dots,g_N$. When $N=2$, we are able to answer this question.

THEOREM 1.7. *Let $N=2$. In order that each function $f(x)$ can be represented in the form* (1.13), *it is necessary and sufficient that X contain no closed lightning bolts.*

PROOF. *Necessity.* If $[x_1,\dots,x_n]$ is a closed lightning bolt, then a straightforward calculation shows that every $f(x)$ representable by (1.13) satisfies

$$f(x_1)-f(x_2)+f(x_3)-f(x_4)+\cdots+f(x_{2m-1})-f(x_{2m})=0.$$

But a function $F(x)$ such that $F(x_1)\neq 0$, $F(x_2)=\cdots=F(x_{2m})=0$ violates this condition, and therefore cannot be represented in the form (1.13).

Sufficiency. The absence of closed lightning bolts guarantees that if points x and y in X can be joined by a lightning bolt, it is unique. This allows one to argue in a similar manner as in the proof of part 1 of Theorem 8.3 in Chapter 1. Now the situation is even simpler, since we have to construct only two functions h_1 and h_2, not three as there, and we need not worry about the boundedness of those functions.

From the examples below (see §2), it follows that a bounded function $f(x)$ can indeed be represented by unbounded functions.

§2. Closedness of the subspace D

We give two forms of the answer to the question raised in §1 regarding the conditions implying that the subspace D (1.7) is closed.

THEOREM 2.1 (Marshall and O'Farrell [**99**]). *For the subspace* (1.7) D *to be closed in* $C(Q)$, *it is necessary and sufficient that there exist a constant* $c>0$ *such that*

$$\sup_{p\in Q}\operatorname*{var}_{R_3(P)} f\le c\sup_{p\in Q}\operatorname*{var}_{R_2(P)} f \quad \textit{for each } f\in A_1. \tag{2.1}$$

Here and below, $\operatorname{var}_E f$ on a set E means the oscillation of f on E.

THEOREM 2.2 (Medvedev [**102**]). *In order that the subspace* (1.7) D *be closed in* $C(Q)$, *it is necessary and sufficient that the lengths of all irreducible lightning bolts contained in* Q *be uniformly bounded.*

In Theorem 2.2 closedness of D is encoded in geometric properties of Q, while in Theorem 2.1 there are two parts: a geometric one associated with the structure of $R_2(P)$ and $R_3(P)$, and an analytic one, verification of (2.1). We shall need some auxiliary considerations.

Best approximation of a bounded set (of values) by a continuous function. As long as we are going to study the problem of best approximation of a function $f(x,y)$ by the sums $\varphi(x)+\psi(y)$, it is natural to start out with the problem of approximation of $f(x,y)$ by functions of one variable only. In a slightly more general form, the problem is the following.

Let T be a compact, and let each point $t\in T$ correspond to a set of real numbers $f(T)$ (there is a multivalued function defined on T). Assume the sets $f(t)$ are uniformly bounded: there exists M such that $|f(t)|\le M$ for all $t\in T$. The problem is to find

$$e(f)=\inf_{\varphi\in C(T)}\sup_{t\in T}\sup_{f(t)}|f(t)-\varphi(t)|, \tag{2.2}$$

and to determine the function $\varphi^\star(t)\in C(T)$ for which the infimum in (2.2) is attained (provided that such a function exists). First, we introduce the functions

$$M(t)=\sup f(t),\qquad m(t)=\inf f(t). \tag{2.3}$$

Consider $t_0\in T$, and let $\Sigma(t_0)=\{\sigma(t_0)\}$ be a directed set of neighborhoods $\sigma(t_0)$ of that point partially ordered with respect to inverse inclusion. Define scalar nets with the index set $\Sigma(t_0)$:

$$\overline{f}(\sigma)=\sup_{t\in\sigma(t_0)}M(t),\qquad \underline{f}(t)=\inf_{t\in\sigma(t_0)}m(t) \tag{2.4}$$

and set

$$M^\star(t_0) = \lim_{\Sigma(t_0)} \overline{f}(\sigma), \qquad m^\star(t_0) = \lim_{\Sigma(t_0)} \underline{f}(\sigma). \tag{2.5}$$

The existence of the limits in (2.5) follows from the monotonicity of $\overline{f}(\sigma)$ and $\underline{f}(\sigma)$. Clearly, $M^\star(t_0) \geq M(t_0)$ and $m^\star(t_0) \leq m(t_0)$.

PROPOSITION 2.3. *We have*

$$e(f) = \frac{1}{2} \sup_{t \in T} \left(M^\star(t) - m^\star(t)\right). \tag{2.6}$$

Moreover, there exists a point $t_0 \in T$ where the supremum in (2.6) *is attained. There also exists a function $\varphi^\star(t) \in C(T)$ for which the infimum* (2.2) *is attained.*

PROOF. The function $M^\star(t)$ is upper semi-continuous, whereas $m^\star(t)$ is lower semi-continuous. Hence, $M^\star(t) - m^\star(t)$ is upper semi-continuous and attains its supremum at a point $t_0 \in T$. Let $\varphi(t) \in C(T)$. To fix the ideas, assume that $\varphi(t_0) \leq \frac{1}{2}\left(M^\star(t_0) + m^\star(t_0)\right)$. Taking $\varepsilon > 0$, consider a neighborhood $\sigma(t_0)$ such that $|\varphi(t) - \varphi(t_0)| < \varepsilon$, $t \in \sigma(t_0)$. In this neighborhood there exists a point t' such that $M(t') > M^\star(t_0) - \varepsilon$. Therefore,

$$\begin{aligned} M(t') - \varphi(t') &\geq M^\star(t) - \varphi(t_0) - 2\varepsilon \geq M^\star(t_0) - \frac{1}{2}\left(M^\star(t_0) + m^\star(t_0)\right) - 2\varepsilon \\ &= \frac{1}{2}\left(M^\star(t) - m^\star(t_0)\right) - 2\varepsilon. \end{aligned}$$

Thus,

$$\sup_{t \in T} \sup_{f(t)} |f(t) - \varphi(t)| \geq \frac{1}{2}\left(M^\star(t_0) - m^\star(t_0)\right). \tag{2.7}$$

Since $\varphi \in C(T)$ is arbitrary, we find that

$$e(f) \geq \frac{1}{2}\left(M^\star(t_0) - m^\star(t_0)\right) = \frac{1}{2} \sup_{t \in T} \left(M^\star(t) - m^\star(t)\right). \tag{2.8}$$

Set $q = M^\star(t_0) - m^\star(t_0)$ and try to construct a continuous function $\varphi^\star(t)$ satisfying the inequalities

$$-\frac{q}{2} + M^\star(t) \leq \varphi^\star(t) \leq \frac{q}{2} + m^\star(t). \tag{2.9}$$

The left-hand side in (2.9) is indeed less than or equal to the right. On the left-hand side of (2.9) we have an upper semi-continuous function, while on the right we have a lower semi-continuous function. According to the well-known theorem of Katětov [**76**] such a continuous function $\varphi^\star(t)$ exists (in [**111,** Chapter XV, Section 4] the construction is given when T is an interval on the real axis, but it extends *mutatis mutandis* to the general case). From (2.9) we obtain

$$\begin{gathered} M(t) - \varphi^\star(t) \leq M^\star(t) - \varphi^\star(t) \leq \frac{q}{2}, \\ \varphi^\star(t) - m(t) \leq \varphi^\star(t) - m^\star \leq \frac{q}{2}. \end{gathered}$$

Therefore,

$$|M(t) - \varphi^\star(t)| \leq \frac{q}{2}, \qquad |\varphi^\star(t) - m(t)| \leq \frac{q}{2}. \tag{2.10}$$

But (2.10) implies that for all $t \in T$,

$$\sup_{f(t)} |f(t) - \varphi^\star(t)| \leq \frac{q}{2},$$

and hence

$$\sup_t \sup_{f(t)} |f(t) - \varphi^\star(t)| \leq \frac{q}{2}. \tag{2.11}$$

Together with (2.8), (2.11) implies (2.6) and all the statements of our proposition.

Proposition 2.3 is proved in [**79**]. A special case when $f(t)$ is a single-valued (discontinuous) function was treated in [**90**].

Best approximation of a bounded set of values by a bounded function. Change the problem now by considering approximation by bounded functions. Now let T be an arbitrary set and let $B(T)$ be the space of bounded functions on T with the usual uniform norm. Find

$$\varepsilon(f) = \inf_{\varphi \in B(T)} \sup_{t \in T} \sup_{f(t)} |f(t) - \varphi(t)|. \tag{2.12}$$

Unlike Proposition 2.3, the following assertion can be viewed as obvious.

Proposition 2.4. *We have*

$$\varepsilon(f) = \frac{1}{2} \sup_{t \in T} (M(t) - m(t)). \tag{2.13}$$

There exist functions $\varphi^\star(t) \in B(T)$ for which the infimum in (2.12) *is attained. One of them is the function*

$$\varphi^\star(t) = \frac{1}{2}(M(t) + m(t)). \tag{2.14}$$

In some sense, the function (2.14) is the best of the best approximations, since its values are optimal for each cross-section of $f(t)$. This need not be true for a solution of the problem (2.12): it is easy to see that the values (2.14) are prescribed only for those cross-sections $f(t_0)$ for which $M(t_0) - m(t_0) = \sup_t (M(t) - m(t))$. For other values of t, functions giving best approximation may diverge somewhat from the function (2.14).

Best approximation of a function of two variables by functions of one variable. Let $Q \subset X \times Y$ be a compact set, where X, Y are compact spaces and $f(x,y) \in C(Q)$. The following problem is a special case of the problem (1.15):

$$e_x(f) = \inf_{\varphi \in C(X)} \|f(x,y) - \varphi(x)\|_{C(Q)} = \inf_{\varphi \in C(X)} \max_x \max_y |f(x,y) - \varphi(x)| \tag{2.15}$$

and, similarly, the problem

$$e_y(f) = \inf_{\psi \in C(Y)} \|f(x,y) - \psi(y)\|_{C(Q)} = \inf_{\psi \in C(Y)} \max_y \max_x |f(x,y) - \psi(y)| \tag{2.15$'$}$$

of approximating $f(x,y)$ by functions $\psi(y) \in C(Y)$. For the problem (2.15) we set

$$M(x) = \max_y f(x,y), \qquad m(x) = \min_y f(x,y),$$
$$\overline{f}(\sigma) = \sup_{x\in\sigma(x_0)} M(x), \qquad \underline{f}(\sigma) = \inf_{x\in\sigma(x_0)} m(x),$$
$$M^\star(x_0) = \lim_{\Sigma(x_0)} \overline{f}(\sigma), \qquad m^\star(x_0) = \lim_{\Sigma(x_0)} \underline{f}(\sigma),$$

$$e_x(f) = \frac{1}{2}\max_{x\in X}\left(M^\star(x) - m^\star(x)\right). \tag{2.16}$$

We leave to the reader the task of writing out the corresponding formulas for the problem (2.15′).

If $Q \subset X\times Y$, X, Y being arbitrary sets, then similar problems rise in the class $B(Q)$:

$$\varepsilon_x(f) = \inf_{\varphi\in B(X)} \|f(x,y) - \varphi(x)\|_{B(Q)} = \inf_{\varphi\in B(X)} \sup_x \sup_y |f(x,y) - \varphi(x)| \tag{2.17}$$

and a similar problem for $\varepsilon_y(f)$. A solution of the problem (2.17) is given by

$$M(x) = \sup_{(x,y)\in Q} f(x,y), \qquad m(x) = \inf_{(x,y)\in Q} f(x,y),$$

$$\varepsilon_x(f) = \frac{1}{2}\sup_{x\in X}\left(M(x) - m(x)\right), \tag{2.18}$$

$$\varphi^\star(x) = \frac{1}{2}\left(M(x) + m(x)\right) \tag{2.19}$$

and by similar formulas for approximation by functions $\psi(y) \in B(Y)$.

If $f(t) \notin C(T)$ in the problem (2.2), then $\varepsilon(f) \le e(f)$ always, and it is easy to find examples when the inequality is strict. However, if $M(t)$ and $m(t)$ are continuous, then $\varepsilon(f) = e(f)$ and the function (2.14) is a best approximation. Hence, in that case enlarging the class of approximating functions does not improve the result.

COROLLARY 2.5. *If X and Y are compact spaces, $Q = X\times Y$, $f \in C(Q)$, then*

$$\varepsilon_x(f) = e_x(f) = \frac{1}{2}\max\left(M(x) - m(x)\right). \tag{2.20}$$

The function (2.19) is a best approximation. Similar statements hold for ε_y, e_y, and $\psi^\star(y)$.

The distance to a subalgebra. The following situation is an important case when $e(f)$ and $\varepsilon(f)$ coincide.

LEMMA 2.6. *Let Q be a compact set, A a closed subalgebra in $C(Q)$ containing constants, and r the equivalence relation on Q defined by A. Then for all $f \in C(Q)$ we have*

$$\begin{aligned}\rho(f,A) &= \inf_{\varphi\in A}\|f-\varphi\| = \inf_{\varphi\in A}\max_{p\in Q}|f(p)-\varphi(p)| \\ &= \frac{1}{2}\max_{p\in Q}\left[\max_{q\in r(p)} f(q) - \min_{q\in r(p)} f(q)\right] = \frac{1}{2}\max_{p\in Q}\operatorname*{var}_{r(p)} f.\end{aligned} \tag{2.21}$$

PROOF. Consider the quotient space $T=Q/r$ whose elements are equivalence classes $t=r(p)$, and the natural projection $\pi: Q\to Q/r$. The space T is compact, and according to the Stone–Weierstrass theorem A can be identified with $C(T)$. The approximation problem (2.21) is then the problem of best approximation of sets of values $f(t)$ assumed by the function $f(p)$ on equivalence classes $t=r(p)$ into which r divides Q. Let us show that for all $t\in T$ we have $M^\star(t_0)=M(t_0)$ and $m^\star(t_0)=m(t_0)$. In view of Propositions 2.3 and 2.4, this will prove the lemma.

Let $\Sigma(t_0)=\{\sigma(t_0)\}$ be a directed set of neighborhoods of a point t_0. For any $\sigma(t_0)\in\Sigma(t_0)$ we have $\overline{f}(\sigma)\geq M(t_0)$. Given an arbitrary $\varepsilon>0$, then for any $\sigma=\sigma(t_0)$ there is a $t_\sigma\in\sigma$ such that $M(t_\sigma)>\overline{f}(\sigma)-\varepsilon\geq M^\star(t_0)-\varepsilon$. Hence, there exists a point $p_\sigma\in\pi^{-1}(t_\sigma)$ such that $f(p_\sigma)\geq M^\star(t_0)-\varepsilon$. From the net $\{p_\sigma\}$ choose a subnet $\{p_\beta\}$ that converges in Q to q_0. Then $\pi(q_0)=\lim\pi(p_\beta)=\lim\pi(p_\sigma)=t_0$. Therefore, $q_0\in\pi^{-1}(t_0)=r(p_0)$. Then $M(t_0)=\sup_{q\in r(p_0)}f(q)\geq f(q_0)=\lim f(p_\beta)\geq M^\star(t_0)-\varepsilon$. Since $M(t_0)\leq M^\star(t_0)$ always, we obtain the equality $M(t_0)=M^\star(t_0)$. The proof that $m(t_0)=m^\star(t_0)$ is the same. The lemma is proved.

Now we can finally start proving the main theorems.

PROOF OF THEOREM 2.1 (about closedness of $D(Q)$). Consider the isomorphism of the quotient algebras

$$A_1/A_3=A_1/A_1\cap A_2\to A_1+A_2/A_2\subset C(Q)/A_2. \tag{2.22}$$

Let $f\in A_1$. Consider the equivalence classes $F=f+A_1\cap A_2$ and $\Phi=f+A_2$ determined by an element f; they are elements in $A_1/A_1\cap A_2$ and A_1+A_2/A_2, respectively. These classes correspond to each other under the isomorphism (2.22). The respective norms (quotient norms) are

$$\begin{aligned}\|F\|_1&=\inf_{\varphi\in A_1\cap A_2}\|f-\varphi\|=\rho(f,A_3), \qquad A_3=A_1\cap A_2,\\ \|\Phi\|_2&=\inf_{\varphi\in A_2}\|f-\varphi\|=\rho(f,A_2).\end{aligned} \tag{2.23}$$

Here ρ is the distance from f to A_3, or A_2, in $C(Q)$. Clearly, $\|\Phi\|_2\leq\|F\|_1$, and the linear operator $F\to\Phi$ defined by (2.22) is continuous. If there is a c such that $\|F\|_1\leq c\|\Phi\|_2$, then A_1+A_2/A_2 is a Banach space. Conversely, if A_1+A_2/A_2 is a Banach space, then according to the Banach Open Mapping Theorem there exists c such that $\|F\|_1\leq c\|\Phi\|_2$. On the other hand, A_1+A_2/A_2 being a Banach space is equivalent to closedness of A_1+A_2. Thus, the inequality

$$\|F\|_1\leq c\|\Phi\|_2 \tag{2.24}$$

with a certain constant c is equivalent to closedness of A_1+A_2. Applying to $\|F\|_1=\rho(f,A_3)$ and $\|\Phi\|_2=\rho(f,A_2)$ formula (2.21) of Lemma 2.6, we complete the proof.

PROOF OF THEOREM 2.2 (about closedness of D). *Necessity.* Assume that the sum A_1+A_2 (i.e., the subspace D) is closed, but still Q contains irreducible lightning bolts of an arbitrarily large length. Then, for each natural number N there exists an irreducible lightning bolt

$$p_0R_1p_1R_2p_2\dots p_{2n+1}R_2p_{2n+2}. \tag{2.25}$$

(We have specified explicitly the relations between the points of the lightning bolt $[p_0p_1\dots p_{2n+2}]$.) Introduce an operator γ by setting for any set $E\subset Q$

$$\gamma(E)=R_1\left(R_2(E)\right). \tag{2.26}$$

The powers of γ are defined accordingly. Set

$$B_k=\gamma^{n-k}\left(p_{2n+2}\right),\qquad k=0,1,\dots,n.$$

Here, $\gamma^0\left(p_{2n+2}\right)=\{p_{2n+2}\}$. The set $B_{n-1}=\gamma\left(p_{2n+2}\right)$ contains points p_{2n+2}, $p_{2n+1}\in R_2\left(p_{2n+2}\right)$, $p_{2n}\in R_1\left(p_{2n+1}\right)\subset R_1\left(R_2\left(p_{2n+2}\right)\right)$. At the same time, B_{n-1} does not contain any other points of (2.25) since it is irreducible. (If a point p_j, $j<2n$ belongs to $\gamma\left(p_{2n+2}\right)$, then we can find a point q such that

$$p_jR_1qR_2p_{2n+2}.$$

But this means that there exists a lightning bolt shorter than (2.25) with end-points p_0 and p_{2n+2}.) In general, B_k contains points $p_{2k+2},p_{2k+3},\dots,p_{2n+2}$ and does not contain points $p_0,p_1,\dots,p_{2k+1}$. In particular, B_0 contains all of (2.25) except p_0 and p_1. Each of the sets B_k is closed (since R_j is the saturation of a closed set, therefore closed) and is saturated with respect to R_1. Set $C_k=Q\backslash B_k$. C_k is an open set and is also saturated with respect to R_1. Since $B_0\supset B_1\supset\cdots\supset B_n$, we have

$$C_0\subset C_1\subset\cdots\subset C_n. \tag{2.27}$$

Also,

$$\{p_0,p_1,\dots,p_{2k+1}\}\subset C_k,\qquad \{p_{2k+2},\dots,p_{2n+2}\}\cap C_k=\emptyset. \tag{2.28}$$

Let us show that

$$\gamma\left(C_k\right)\subset C_{k+1},\qquad k=0,\dots,n-1. \tag{2.29}$$

Indeed, if the inclusion (2.29) fails for some k, then there exists $y_1\in C_k$ such that $y_1R_2y_2R_1y_3$, where $y_3\notin C_{k+1}$. This implies that $y_3\in B_{k+1}=Q\backslash C_{k+1}$. Then $y_2\in B_{k+1}$, since B_{k+1} is saturated with respect to R_1. But then $y_1\in R_2\left(B_{k+1}\right)\subset\gamma\left(B_{k+1}\right)=B_k$, and this contradicts the condition $y_1\in C_k$.

In view of (2.27) and (2.28) we have

$$\pi_1\left(p_0\right)=\pi_1\left(p_1\right)\subset\pi_1\left(C_0\right)\subset\pi_1\left(C_1\right)\subset\cdots\subset\pi_1\left(C_n\right). \tag{2.30}$$

All the sets $\pi_1\left(C_k\right)$, $k=0,\dots,n$, are open. Indeed, the sets $\pi_1\left(C_k\right)$ and $\pi_1\left(B_k\right)$ are disjoint, since otherwise the sets $\pi_1^{-1}\left(\pi_1\left(C_k\right)\right)=C_k$ and $\pi_1^{-1}\left(\pi_1\left(B_k\right)\right)=B_k$ are not disjoint either. (The equality $\pi_1^{-1}\left(\pi_1(E)\right)=E$ means that E is saturated with respect to R_1.) Since $\pi_1\left(B_k\right)$ is closed, $\pi_1\left(C_k\right)=X-\pi_1\left(B_k\right)$ is open.

Set $F_0=\pi_1\left(p_0\right)$. It is a closed set (a point). Since X is compact, there exists an open set V_0 such that

$$F_0\subset V_0\subset\overline{V}_0\subset\pi_1\left(C_0\right)$$

(this is the so-called small Urysohn lemma, cf. [**1**], [**117**]). From $\pi_1^{-1}\left(\overline{V}_0\right)\subset\pi_1^{-1}\left(\pi_1\left(C_0\right)\right)=C_0$, it follows that $\gamma\left(\pi_1^{-1}\left(\overline{V}_0\right)\right)\subset C_1$. Set $F_1=\pi_1\left[\gamma\left(\pi_1^{-1}\left(\overline{V}_0\right)\right)\right]$. Then F_1 is a closed set, and $F_1\subset\pi_1\left(C_1\right)$. Again, we can find an open set V_1 such that $F_1\subset V_1\subset\overline{V}_1\subset\pi_1\left(C_1\right)$. Set $F_2=\pi_1\left[\gamma\left(\pi_1^{-1}\left(\overline{V}_1\right)\right)\right]$. F_2 is closed, and since $\pi_1^{-1}\left(\overline{V}_1\right)\subset C_1$, then $\gamma\left(\pi_1^{-1}\left(\overline{V}_1\right)\right)\subset\gamma\left(C_1\right)\subset C_2$. Hence, $F_2=\pi_1\left(C_2\right)$. Let us find an open set V_2 with $F_2 \subset V_2 \subset \overline{V}_2 \subset \pi_1(C_2)$, and let $F_3 = \pi_1\left[\gamma\left(\pi_1^{-1}\left(\overline{V}_2\right)\right)\right]$, etc. We obtain a sequence $V_0, V_1, \dots, V_n$ of open sets in X and a sequence of closed sets such that

$$\begin{gathered} F_0 \subset V_0 \subset F_1 \subset V_1 \subset F_2 \subset V_2 \subset \cdots \subset V_{n-1} \subset F_n \subset \pi_1\left(C_n\right), \\ F_k = \pi_1\left[\gamma\left(\pi_1^{-1}\left(\overline{V_{k-1}}\right)\right)\right], \qquad F_k \subset \pi_1\left(C_k\right). \end{gathered} \tag{2.31}$$

We have

$$\pi_1^{-1}\left(F_k\right) \supset \gamma\left(\pi_1^{-1}\left(\overline{V_{k-1}}\right)\right) \subset \gamma\left(\pi_1^{-1}\left(F_{k-1}\right)\right), \qquad k=1,\dots,n.$$

Hence,

$$\pi_1^{-1}\left(F_k\right) \supset \gamma^k\left(\pi_1^{-1}\left(F_0\right)\right) \supset \left\{p_0, \dots, p_{2k+1}\right\}. \tag{2.32}$$

On the other hand, in view of (2.28) and (2.31)

$$\pi_1^{-1}\left(F_{k-1}\right) \cap \left\{p_{2k}, \dots, p_{2n+2}\right\} \subset C_{k-1} \cap \left\{p_{2k}, \dots, p_{2n+2}\right\} = \emptyset. \tag{2.33}$$

Therefore

$$\pi_1^{-1}\left(F_k\right) \backslash \pi_1^{-1}\left(F_{k-1}\right) \supset \left\{p_{2k}, p_{2k+1}\right\} \neq \emptyset, \tag{2.34}$$

whence

$$F_k \backslash V_{k-2} \subset F_k \backslash F_{k-1} \neq \emptyset. \tag{2.35}$$

In addition, $F_k \backslash V_{k-2}$ is a closed set. For the sake of uniformity, complete our notation by setting

$$V_{-1} = \emptyset, \qquad V_n = \pi_1\left(C_n\right), \qquad F_{n+1} = X. \tag{2.36}$$

By Urysohn's lemma we can construct a continuous real-valued function $h_k(x)$, $k = 1, 2, \dots, n+1$, satisfying the following conditions:

$$\begin{gathered} h_k(x) = k-1, \quad \text{for } x \in F_{k-1} \backslash V_{k-2}; \\ h_k(x) = k, \quad \text{for } x \in F_k \backslash V_{k-1}; \\ k-1 \leq h_k(x) \leq k \quad \text{for all } x \in F_k \backslash V_{k-2}. \end{gathered} \tag{2.37}$$

It is possible for the set $F_{k-1} \backslash V_{k-2}$ or $F_k \backslash V_{k-1}$ (or even both of them) to be empty. The domains of definition of the functions h_k may intersect only for consecutive indices, yet on those intersections the functions h_k and h_{k-1}, or h_k and h_{k+1}, as defined by (2.37), coincide.

The function $h(x)$ defined by

$$h(x) = h_k(x) \quad \text{for } x \in F_k \backslash V_{k-2}, \qquad k = 1, \dots, n+1 \tag{2.38}$$

is well defined on all of X and is continuous. Set $f = h \circ \pi_1$. Then $F \in A_1$. We have $\pi_1\left(p_0\right) = F_0 = F_0 \backslash V_{-1}$. Therefore, $h \circ \pi_1\left(p_0\right) = 0$. Moreover, for $k = n+1$ the only point remaining in (2.34) is p_{2n+2}; hence $\pi_1\left(p_{2n+1}\right) \in F_{n+1} \backslash V_{n-1}$. Therefore, $f\left(p_{2n+2}\right) = h \circ \pi\left(p_{2n+2}\right) = n+1$. Comparing the relations R and R_3, we see that $R_3(p) \supset R(p)$ always. Hence,

$$\sup_{p \in Q} \underset{R_3(p)}{\operatorname{var}} f \geq \sup_{p \in Q} \underset{R(p)}{\operatorname{var}} \geq \underset{R(p_0)}{\operatorname{var}} f \geq n+1. \tag{2.39}$$

(Incidentally, it is not hard to see that the last inequality in (2.39) is, in fact, an equality.) Now let E be an equivalence class with respect to R_2. Let k be the smallest index for which $E \cap \pi_1^{-1}(F_k) \neq \emptyset$. ($\pi_1^{-1}(F_{n+1}) = \pi_1^{-1}(X) = Q$.) If $0 < k < n+1$, then $E \cap \pi_1^{-1}(F_{k-1}) = \emptyset$. But $\pi_1^{-1}(F_{k+1}) = \gamma\left(\pi_1^{-1}\left(\overline{V}_k\right)\right) \supseteq \gamma\left(\pi_1^{-1}(F_k)\right) \supset R_2\left(\pi_1^{-1}(F_k)\right) \supset E$. Therefore,

$$E \subset \pi_1^{-1}(F_{k+1} \backslash F_k) \subset \pi_1^{-1}(F_{k+1} \backslash V_{k-2}).$$

Hence $\operatorname{var}_E f \leq 2$. For $k = 0$ or $k = n+1$, we obtain similarly that $\operatorname{var}_E f \leq 1$. Hence

$$\sup_{p \in Q} \operatorname{var}_E f \leq 2. \tag{2.40}$$

Comparing (2.39) and (2.40), we see that the condition (2.1) in the Marshall–O'Farrell criterion is violated. Necessity in Theorem 2.2 is proved.

Sufficiency. We have to show that the subspace D is closed provided that the lengths of all irreducible lightning bolts in Q are bounded by a number N. As was noted earlier, for any $p \in Q$ the inclusion $R_3(p) \supseteq R(p)$ always holds, and since $R_3(p)$ is closed $R_3(p) \supseteq \overline{R(p)}$. Let us show that in our case $R_3(p) = R(p)$. If $q \in R(p)$, then among the lightning bolts joining p to q we can choose one that has the smallest number of vertices. This lightning bolt is going to be irreducible, and so its length is at most N. Thus, any two points in a equivalence class with respect to R can be joined by a lightning bolt that has at most N vertices. This immediately implies that in any case

$$R(p) = \gamma^N(p), \tag{2.41}$$

where γ is the operator defined by (2.26). But $\gamma^N(p)$ is closed; hence the equivalence class $R(p)$ is a closed set.

Similarly, if F is a closed subset of Q, then $R(F) = \gamma^N(F)$, and this implies that $R(F)$ is also closed. Therefore (see [**117**]), Q/R is a normal compact space and the natural projection $\pi : Q \to Q/R$ is a closed mapping. For two distinct equivalence classes $a = R(p_1)$ and $b = R(p_2)$, we can construct a function Φ continuous on Q such that $\Phi(a) = 1$, $\Phi(b) = 0$. The function $F(p) = \Phi \circ \pi(p)$ is continuous on Q. Since it is constant on every equivalence class $R(p)$, $F \in A_3$. At the same time it takes distinct values on $R(p_1)$ and $R(p_2)$, which is only possible if p_1 is not equivalent to p_2 with respect to R. Thus, $R_3(p) = R(p)$.

Now, let $f \in A_1$. Choose two points p' and p'' in $R_3(p) = R(p)$. Then p' and p'' can be joined by a lightning bolt $[p', p_1, \dots, p_k, p'']$, and $k \leq N-2$. To fix the ideas assume that $p' R_1 p_1, p_R R_2 p''$ (all the remaining possibilities are treated similarly). Then $p_1 R_2 p_2, p_3 R_2 p_4, \dots, p_{k-2} R_2 p_{k-1}, p_k R_2 p''$, and we have

$$\begin{aligned} |f(p') - f(p'')| &\leq |f(p_1) - f(p_2)| + |f(p_3) - f(p_4)| + \cdots \\ &+ |f(p_k) - f(p'')| \leq \left[\frac{N}{2}\right] \sup_{p \in Q} \operatorname*{var}_{R_2(p)} f \end{aligned} \tag{2.42}$$

($f(p') = f(p_1), f(p_2) = f(p_3), \cdots, f(p_{k-3}) = f(p_{k-2}), f(p_{k-1}) = f(p_k)$, since $f \in A_1$). From (2.42) we obtain

$$\sup_{p \in Q} \operatorname*{var}_{R_3(p)} f \leq \left[\frac{N}{2}\right] \sup_{p \in Q} \operatorname*{var}_{R_2(p)} f;$$

hence, in view of Theorem 2.1 (where we have to set $c = [N/2]$), the subspace D is closed. The proof is now complete.

Bars and a cross. Let Q, X, Y be sets, and $\pi_1 : Q \to X$, $\pi_2 : Q \to Y$ surjective mappings. If for $p_0 \in Q$ we have

$$\pi_2\left[\pi_1^{-1}\left(\pi_1\left(p_0\right)\right)\right]=Y \tag{2.43}$$

or, using different notation,

$$\pi_2\left[R_1\left(p_0\right)\right]=Y, \tag{2.43$'$}$$

then $R_1(p_0)$ is called a y-bar passing through p_0. Similarly, if

$$\pi_1\left(\pi_2^{-1}\left(\pi_2\left(p_0\right)\right)\right]=\pi_1\left(R_2\left(p_0\right)\right)=X, \tag{2.44}$$

then $R_2(p_0)$ is an x-bar passing through p_0. In the case when $Q \subset X \times Y$ existence of a y-bar through p_0 means that $\pi_1(p_0) \times Y \subset Q$. Similarly, existence of an x-bar means that $\pi_2(p_0) \times X \subset Q$. If $R_1(p_0)$ is a y-bar and $R_2(p_0)$ is an x-bar, then $R_1(p_0) \cup R_2(p_0)$ is called a cross (passing through p_0). This notion was used, e.g., in [**113**], [**109**], [**110**], [**81**].

COROLLARY 2.7. *If Q, X, Y are compact sets and Q contains at least one bar, then $D(Q)$ is closed in $C(Q)$.*

PROOF. Let Q contain an x-bar through p_0 and let $p_1 \in Q, p_2 \in Q$ be two points. By the hypothesis there exist points $q_1 \in R(p_0)$ and $q_2 \in R(p_0)$ such that $\pi_1(p_1) = \pi_1(q_1)$ and $\pi_1(p_2) = \pi_1(q_2)$. Then, $[p_1 R_1 q_1 R_2 q_2 R_1 p_2]$ is a lightning bolt.

EXAMPLES. Let us give few examples illustrating the theory presented here. All examples are constructed for $Q \subset \mathbb{R}^2$ with X, Y being projections of Q on the coordinate axes; the subspace D has the form (1.8), $A_1 = \{\varphi(x)\}$, $A_2 = \{\psi(y)\}$, $A_3 = A_1 \cap A_2$. For a more complete analysis of the examples below let us note the following fact (cf. Theorem 4.15 below). In order that $\overline{D(Q)} = C(Q)$, it is necessary that there be no closed lightning bolts in Q. (This is easily verified by arguments in the proof of necessity in Theorem 1.7.) With the same goal in mind, we shall be appealing to Lemma 4.1 in order to check that a measure μ orthogonal to $D(Q)$ is a zero measure.

EXAMPLE 2.8. A_3 does not consist of constants only. Let

$$\begin{aligned} Q_1 &= \{(x,y), y=x, 0 \le x \le 1\}, \\ Q_2 &= \{(x,1), 1 \le x \le 2\}, \\ Q_3 &= \{(2,y), 1 \le y \le 2\}, \\ Q &= Q_1 \cup Q_2 \cup Q_3. \end{aligned}$$

Obviously, $A_1 \neq A_2 \neq A_3$, and A_3 contains some other functions besides constants. For instance,

$$f(x,y)=\begin{cases} x, & 0 \le x \le 1, \\ 1, & 1 \le x \le 2, \end{cases}$$

can also be written as

$$f(x,y)=\begin{cases} y, & 0 \le y \le 1, \\ 1, & 1 \le y \le 2; \end{cases}$$

hence $f(x,y) \in A_3$ although $f(x,y) \neq \text{const}$.

EXAMPLE 2.9. Here we illustrate situations when $D = C(Q)$ and $D \neq C(Q)$.

(a) $Q = \{(0,0),(1,0),(1,1),(0,1)\}$. $D \neq C(Q)$ because Q is a closed lightning bolt.

(b) $Q = \{(0,0),(1,0),(1,2),(2,2),(2,0)\}$. $D \neq C(Q)$ because Q contains a closed lightning bolt.

(c) $Q = \{(0,0),(1,0),(1,2),(2,2)\}$. $D = C(Q)$.

In examples (a), (b), (c), $D = \overline{D}$. In general, when Q is a finite set, the assumptions of Theorem 2.2 are trivially satisfied and $\overline{D} = D$ (here, $C(Q)$ is a finite-dimensional space and D, of course, is also finite-dimensional, so automatically $\overline{D} = D$).

EXAMPLE 2.10. Here we illustrate cases when $D = \overline{D}$ and $D \neq \overline{D}$.

(a) Q is a union of two parallel segments that are not parallel to the coordinate axes. $D = C(Q) = \overline{D}$. (The assumptions of Theorem 1.6 are satisfied.)

(b) Three such segments. By Theorem 2.2, $D = \overline{D}$. If the segments are sufficiently long, then $D \neq C(Q)$ (there is a closed lightning bolt in Q).

(c) Q is a boundary of the triangle with vertices $(0,0)$, $(1/2,0)$, $(1,1)$. Here, $D \neq C(Q)$ (there exist arbitrarily long irreducible lightning bolts). One can show (see Lemma 4.1 below) that $\overline{D} = C(Q)$ and, therefore, $D \neq \overline{D}$. (The latter follows from Theorem 2.2.)

(d) Q consists of a polygonal path whose sides are parallel to the coordinate axes and whose vertices are

$$(0,0),(1,0),(1,1),\left(1+\frac{1}{2^2},1\right),\left(1+\frac{1}{2^2},1+\frac{1}{2^2}\right),$$
$$\left(1+\frac{1}{2^2}+\frac{1}{3^2},1+\frac{1}{2^2}\right),\left(1+\frac{1}{2^2}+\frac{1}{3^2},1+\frac{1}{2^2}+\frac{1}{3^2}\right),\dots.$$

We add to the path the limit point for the vertices

$$\left(\frac{\pi^2}{6},\frac{\pi^2}{6}\right),\left(\frac{\pi^2}{6}=\sum_{n=1}^{\infty}\frac{1}{n^2}\right).$$

By Lemma 4.1 one can verify that $D = C(Q)$. However, $D \neq \overline{D}$ by Theorem 2.2. Construct a function $f(p) = f(x,y)$ on Q as follows. On the link joining $(0,0)$ to $(1,0)$, $f(x,y)$ is continuously increasing from 0 to 1; on the link from $(1,0)$ to $(1,1)$ it continuously decreases from 1 to 0; on the link from $(1,1)$ to $\left(1+\frac{1}{2^2},1\right)$ it increases from 0 to $\frac{1}{2}$; on the link from $\left(1+\frac{1}{2^2},1\right)$ to $\left(1+\frac{1}{2^2},1+\frac{1}{2^2}\right)$ it decreases from $\frac{1}{2}$ to 0; on the next link it increases from 0 to $\frac{1}{3}$, etc. At the point $\left(\frac{\pi^2}{6},\frac{\pi^2}{6}\right)$ set the value of f equal to 0. Clearly, f is a continuous function and therefore can be uniformly approximated by the sums $\varphi(x)+\psi(y)$, where φ, ψ are continuous. At the same time, $f(x,y)$ can be represented in the form

$$f(x,y) = \Phi(x) + \Psi(y).$$

(In general, according to Theorem 1.7 any function on Q is representable by such a sum.) However, it is easy to see that in any such representation Φ and Ψ cannot be bounded—once more confirming that $\overline{D} \neq D$, although $\overline{D} = C(Q)$. Here, the equivalence class with respect to R_3 of any point is all of Q, it contains $R((0,0))$, and $R\left(\left(\frac{\pi^2}{6},\frac{\pi^2}{6}\right)\right) = \left(\frac{\pi^2}{6},\frac{\pi^2}{6}\right)$.

(e) Add to the set Q in the previous example one more copy of Q adjacent to $\left(\frac{\pi^2}{6},\frac{\pi^2}{6}\right)$ from "above", the set Q_1. Let $\tilde{Q} = Q \cup Q_1$. For all $p \in Q$ we have $R_3(p) = \tilde{Q}$. There are three equivalence classes with respect to R: the point $q = \left(\frac{\pi^2}{6},\frac{\pi^2}{6}\right)$, $Q\backslash q$, and $Q_1\backslash q$.

§3. Proximinality

Propositions 2.3–2.4, Corollary 2.5, Lemma 2.6 all contain statements concerning proximinality of subspaces of functions $g \circ \Phi(p)$, where Φ is a given mapping. In this section we treat problems concerning proximinality of subspaces of sums of two superpositions. First, consider bounded functions. Similarly to (1.8), assume that X, Y and $Q \subset X \times Y$ are arbitrary sets, $\pi_1 : X \times Y \to X$, $\pi_2 : X \times Y \to Y$ are canonical projections, $X = \pi_1(Q)$, $Y = \pi_2(Q)$. Functions $f(p)$, $p \in Q$, can be written as $f(x,y)$, $x \in X$, $y \in Y$.

The subspace $BD = BD(Q)$ in $B(Q)$ consists of functions $\varphi(x) + \psi(y)$:

$$BD = \{\varphi(x) + \psi(y)\}, \qquad \varphi(x) \in B(X), \quad \psi(y) \in B(Y). \tag{3.1}$$

Also, in addition to the subspace BD we consider in $B(Q)$ another subspace

$$\widetilde{BD} = \{\varphi(x) + \psi(y)\}, \qquad \text{for all } \varphi(x), \psi(y) \text{ with } \varphi(x) + \psi(y) \in B(Q). \tag{3.1'}$$

It follows from the example (2.10d) that in general BD and $\widetilde{BD}$ are different.

THEOREM 3.1. *The subspace $\widetilde{BD}$ is always proximinal.*

PROOF. In every equivalence class with respect to the relation R choose a point. Let $R(p_1)$ be such a class in which the point p_1 is chosen. Each point $p \in R(p_1)$ can be joined to p_1 by a finite lightning bolt. Among the lightning bolts joining p and p_1 there exist irreducible ones. Fix one such irreducible lightning bolt $[p_1, p_2, \dots, p_m = p]$. We call it *marked*, and denote it by $\{p_1, p\}$. If $p_i = (x_i, y_i)$, $p_{i+1} = (x_{i+1}, y_{i+1})$, then either $y_i = y_{i+1}$, or $x_i = x_{i+1}$. In the first case we shall write $i \in C_x$, in the second case $i \in C_y$. Consider the following linear functionals on $B(Q)$:

$$\begin{aligned} A_p(f) &= \sum_{\substack{1 \le i \le m-1 \\ i \in C_x}} \left[f(p_{i+1}) - f(p_i)\right]; \\ B_p(f) &= \sum_{\substack{1 \le i \le m-1 \\ i \in C_y}} \left[f(p_{i+1}) - f(p_i)\right]. \end{aligned} \tag{3.2}$$

If $p = p_1$, then both sums are defined to be zero. For all $p \in R(p_1)$ these functionals are continuous in the weak $(\star)$ topology of $B(Q)$ defined by $B(Q) = \ell^1(Q)^\star$. Also, for any two points p and $\tilde{p}$ that belong to the same equivalence class $R(p_1)$ define the linear functionals

$$L_{A,p,\tilde{p}}(f) = (A_{\tilde{p}} - A_p)(f); \quad L_{B,p,\tilde{p}}(f) = (B_{\tilde{p}} - B_p)(f). \tag{3.3}$$

These functionals are also continuous with respect to the weak ($\star$) topology in $B(Q)$. Considering all the equivalence classes with respect to R into which R divides Q, denote by $\mathcal{M}$ the set of those functionals (3.3) for which the following additional requirements hold ($p = (x, y)$, $\tilde{p} = (\tilde{x}, \tilde{y})$):

$$L_{A,p,\tilde{p}}(f) \in \mathcal{M} \quad \text{if } x = \tilde{x}; \qquad L_{B,p,\tilde{p}}(f) \in \mathcal{M} \quad \text{if } y = \overline{y}. \tag{3.4}$$

ASSERTION 3.2. $\widetilde{BD} = \mathcal{M}^{\perp}$.

PROOF. Let $f \in \widetilde{BD}$ and hence $f(p) = \varphi(p) + \psi(p)$, where $\varphi(p) = \varphi(x)$, $\psi(p) = \psi(y)$. If $p \in R(p_1)$ and $\{p_1, p\}$ is the marked lightning bolt, then

$$\psi(p_{i+1}) = \psi(p_i), \ i \in C_x, \qquad \varphi(p_{i+1}) = \varphi(p_i), \ i \in C_y.$$

Therefore,

$$\begin{aligned} \varphi(p_{i+1}) - \varphi(p_i) &= f(p_{i+1}) - f(p_i), \qquad i \in C_x, \\ \psi(p_{i+1}) - \psi(p_i) &= f(p_{i+1}) - f(p_i), \qquad i \in C_y. \end{aligned} \tag{3.5}$$

Hence, using the functionals (3.2), we obtain

$$\varphi(p) - \varphi(p_1) = A_p(f), \qquad \psi(p) - \psi(p_1) = B_p(f). \tag{3.6}$$

If $L_{A,p,\tilde{p}} \in \mathcal{M}$, $p = (x, y)$, $\tilde{p} = (\tilde{x}, \tilde{y})$, then in view of (3.4) $x = \tilde{x}$. Hence p and $\tilde{p}$ belong to the same equivalence class with respect to the relation R. Let it be a class $R(p_1)$. Then from (3.1)–(3.6) it follows that

$$L_{A,p,\tilde{p}}(f) = A_{\tilde{p}}(f) - A_p(f) = \varphi(\tilde{p}) - \varphi(p) = \varphi(\tilde{x}) - \varphi(x) = 0. \tag{3.7}$$

Similarly, for $L_{B,p,\tilde{p}} \in \mathcal{M}$ we obtain

$$L_{B,p,\tilde{p}}(f) = 0. \tag{3.7'}$$

Therefore, $\widetilde{BD} \subset \mathcal{M}^{\perp}$. Now let $f \in B(Q)$ be orthogonal to $\mathcal{M}$. Construct $\varphi(x)$ and $\psi(y)$ such that $f = \varphi(x) + \psi(y)$. Taking an equivalence class $R(p_1)$, set $\varphi(p_1) = f(p_1)$, $\psi(p_1) = 0$ at the marked point p_1. At any $p \in R(p_1)$ define the functions $\varphi(p)$ and $\psi(p)$ by (3.6). Do this for all the equivalence classes. From (3.2) and (3.6) it follows right away that $f(p) = \varphi(p) + \psi(p)$. Now let points $p = (x, y) \in R(p_1)$ and $\tilde{p} = (\tilde{x}, \tilde{y})$ be such that $x = \tilde{x}$. Then, $\tilde{p} \in R(p_1)$ and $L_{A,p,\tilde{p}} \in \mathcal{M}$. Therefore, $L_{A,p,\tilde{p}}(f) = 0$. From (3.6), (3.3) it follows that $\varphi(p) - \varphi(\tilde{p}) = 0$. Thus, $\varphi(p)$ depends only on the coordinate x of a point $p : \varphi(p) = \varphi(x)$. Similarly, one shows that $\psi(p)$ depends only on the y-coordinate. Thus, $f = \varphi(x) + \psi(y)$, and so $\mathcal{M}^{\perp} \subset \widetilde{BD}$. Thus, $\mathcal{M}^{\perp} = \widetilde{BD}$, and Assertion 3.2 is proved.

In order to complete the proof of Theorem 3.1, note that from Assertion 3.2 it follows that $\widetilde{BD}$ is weak ($\star$) closed, and weak ($\star$) closedness implies proximinality.

THEOREM 3.3. *If the lengths of irreducible lightning bolts in the set Q are uniformly bounded, then $BD = \widetilde{BD}$ and hence the subspace $BD(Q)$ is proximinal.*

PROOF. By the hypothesis the number of vertices of any given lightning bolt does not exceed some constant M. Let $f(p) = \varphi(x) + \psi(y) \in \widetilde{BD}$. On each equivalence class $R(p_1)$ functions φ and ψ are defined up to additive constants c_1 and $c_2 = -c_1$, respectively. Therefore, we can assume that at a given point p_1 we have $\varphi(p_1) = f(p_1)$ and $\psi(p_1) = 0$. Then from (3.2) and (3.6) we obtain (since f is a bounded function!)

$$|\varphi(p)| \leq M\|f\|, \qquad |\psi(p)| \leq M\|f\|.$$

Thus, $\varphi \in B(X)$, $\psi \in B(Y)$, and $f \in BD$. Hence $\widetilde{BD} = BD$, and from Theorem 3.1 it follows that BD is proximinal.

COROLLARY 3.4. *If the set Q contains either an x-bar or a y-bar, then BD is proximinal.*

PROOF. As in Corollary 2.7, we conclude that the lengths of all irreducible lightning bolts are bounded by 4.

Now consider conditions that would provide proximinality of the subspace D as in (1.8) in $C(Q)$. We shall only consider compact metric spaces. If V is a metric space, we denote by $\sigma(a,r)$ the closed ball of radius r centered at $a \in V$.

THEOREM 3.5. *Let X, Y, $Q \subset X \times Y$ be compact metric spaces and let a subspace D be defined by formulas* (1.8), *where $\pi_1 : Q \to X$, $\pi_2 : Q \to Y$ are the natural projections (considered on Q only), $\pi_1(Q) = X$, $\pi_2(Q) = Y$. Suppose that for any point $x \in X$ and any $\delta > 0$ there exists $\delta_0 = \delta_0(X)$, $0 < \delta_0 < \delta$, such that the set*

$$\pi_1^{-1}\left(\sigma\left(x, \delta_0\right)\right) \tag{3.8}$$

has a y-bar. Then the subspace D is proximinal in $C(Q)$.

First, we shall establish the following lemma.

LEMMA 3.6. *If the assumptions of Theorem* 3.5 *hold, then the lengths of irreducible lightning bolts in Q are uniformly bounded.*

PROOF. From the compactness of X it follows that there exist a finite number of closed balls $\sigma(x_k, \delta_k)$, $x_k \in X$, $k = 1, \dots, \nu$, covering X, such that each set

$$Q_k = \pi_1^{-1}\left(\sigma\left(x_k, \delta_k\right)\right), \qquad k = 1, \dots, \nu,$$

contains a y-bar. Let $x_k^0 \in \sigma(x_k, \delta_k)$ be such a point corresponding to a y-bar of the set Q_k. If $p = (x, y)$ and $\tilde{p} = (\tilde{x}, \tilde{y})$ are two points from Q_k, the points $\left(x_k^0, y\right)$ and $\left(x_k^0, \tilde{y}\right)$ also lie in Q_k. Therefore, p and $\tilde{p}$ can be joined by a lightning bolt $\left[p, \left(x_k^0, y\right), \left(x_k^0, \tilde{y}\right), \tilde{p}\right]$ of length 4.

Since $\bigcup_{k=1}^{\nu} Q_k = Q$, the number of equivalence classes in Q with respect to the relation R is finite, and each such class is a union of some of the sets Q_k. Take a point $p_k \in Q_k$ in each of the sets Q_k (e.g., $p_k = \left(x_k^0, y_k\right)$). If Q_i and Q_j belong to the same equivalence class, let m_{ij} denote the length of an irreducible lightning bolt joining p_i and p_j. Now, if p and $\tilde{p}$ belong to the same equivalence class with respect to R, where $p \in Q_i$, $\tilde{p} \in Q_j$, then in view of the above argument, the points p, p_i, p_j, $\tilde{p}$ can be included in one lightning bolt whose length in any case does not exceed $4 + m_{ij} + 4$. Hence all irreducible lightning bolts in Q have length bounded by $M + 8$, where $M = \max m_{ij}$.

PROOF OF THEOREM 3.5. Let $f \in C(Q) \subset B(Q)$. In view of Theorem 3.3 and Lemma 3.6, the function $f(p) = f(x, y)$ has the best approximation $\varphi_0(x) + \psi_0(y)$ in the space BD. We shall use the following notation. If V is a metric space, $W \subset V$ is a subset and $v \in V$ an element, then $\operatorname{dist}(v, W)$ is the distance from v to W. Let

$$d = \operatorname{dist}(f, BD) \tag{3.9}$$

in $B(Q)$. Consider the function $F(x, y) = f(x, y) - \psi_0(y)$. Obviously,

$$d = \operatorname{dist}(F, B(X)), \tag{3.10}$$

where $B(X)$ is viewed as a subspace of $B(Q)$. As in Propositions 2.3 and 2.4, introduce for $F(x, y)$ the functions $M(x)$, $m(x)$, $M^\star(x)$, $m^\star(x)$. From Proposition 2.4 and (3.10) it follows that

$$d = \frac{1}{2} \sup_{x \in X} (M(x) - m(x)). \tag{3.11}$$

So, for all $x \in X$, we have

$$M(x) - m(x) \leq 2d. \tag{3.12}$$

Let us also show that

$$M^\star(x) - m^\star(x) \leq 2d \tag{3.13}$$

for all $x \in X$. Take an arbitrary $\varepsilon > 0$. The uniform continuity of $f(x, y)$ on Q implies the existence of $\delta > 0$ such that whenever $p(x, \tilde{x}) < \delta$ and (x, y), $(\tilde{x}, y)$ are points in Q,

$$|f(x, y) - f(\tilde{x}, y)| < \varepsilon. \tag{3.14}$$

Fix $x_0 \in X$ and choose $\delta_1 < \delta$ so that the set $\pi_1^{-1}(\sigma(x_0, \delta_1))$ has a y-bar. Let $x_1 \in \sigma(x_0, \delta_1)$ be a value of x corresponding to that bar. For every $x \in \sigma(x_0, \delta_1)$ we use (3.14) and the existence of a y-bar corresponding to x_1 to get

$$M(x) - \sup_{(x,y) \in Q} F(x, y) \leq \sup_{(x_1,y) \in Q} F(x_1, y) + \varepsilon = M(x_1) + \varepsilon;$$
$$m(x) = \inf_{(x,y) \in Q} F(x, y) \geq \inf_{(x_1,y) \in Q} F(x_1, y) - \varepsilon = m(x_1) - \varepsilon.$$

Therefore,

$$\begin{aligned} M^\star(x_0) &\leq \sup_{x \in \sigma(x_0,\delta)} M(x) \leq M(x_1) + \varepsilon; \\ m^\star(x_0) &\geq \inf_{x \in \sigma(x_0,\delta)} m(x) \geq m(x_1) - \varepsilon. \end{aligned} \tag{3.15}$$

Since $M(x_1) - m(x_1) \leq 2d$ (in view of (3.12)), we obtain from (3.15) that

$$M^\star(x_0) - m^\star(x_0) \leq M(x_0) - m(x_1) + 2\varepsilon \leq 2dt\varepsilon.$$

Since $\varepsilon > 0$ is arbitrary, (3.13) follows.

According to Proposition 2.3,

$$\text{dist}\,(F, C(X)) = \frac{1}{2} \sup_{x \in X} [M^\star(x) - m^\star(x)] \geq \text{dist}\ (F, B(X)) = d. \tag{3.16}$$

Comparing (3.16) and (3.13), we conclude that

$$\text{dist}\,(F, C(X)) = d.$$

Proposition 2.3 also implies that there exists the best approximation $\varphi^\star(x) \in C(X)$ to $F(x, y)$ for which

$$\|F(x, y) - \varphi^\star(x)\| = \|f(x, y) - \varphi^\star(x) - \psi(y)\| = d. \tag{3.17}$$

Consider a continuous function $\Phi(x, y) = f(x, y) - \varphi^\star(x)$. From (3.17) it follows that

$$\text{dist}\,(\Phi, B(Y)) = d. \tag{3.18}$$

Construct for $\Phi(x, y)$ functions $M(y)$ and $m(y)$ similar to the earlier construction of $M(x)$ and $m(x)$. According to Proposition 2.4

$$\text{dist}\,(\Phi, B(Y)) = \frac{1}{2} \sup_{y \in Y} [M(y) - m(y)]\,. \tag{3.19}$$

However, Lemma 2.6 also shows that in this situation Φ is continuous, Q is a compact set and the distance $\text{dist}(\Phi, C(Y))$ is equal to the distance in (3.19). So, again, $M^\star(y) = M(y)$ and $m^\star(y) = m(y)$. From Proposition 2.3 it follows that the best approximation $\psi^\star(y) \in C(Y)$ to Φ does exist. So,

$$\|\Phi - \psi^\star\| = \|f(x, y) - \varphi^\star(x) - \psi^\star(y)\| = d,$$

and hence $\varphi^\star(x) + \psi^\star(y)$ is the best approximation to f in the subspace D. Proximinality of D in $C(Q)$ is proved.

REMARKS 3.7. 1. It is clear from the proof that $\varphi^\star + \psi^\star$ is the best approximation to f not only in $C(Q)$ but also in $B(Q)$.

2. The asymmetry of the assumptions of the theorem with respect to x and y should be stressed: presence of local y-bars suffices, while presence of crosses does not. (Of course, one can interchange the roles of the variables in the statement of the theorem.)

3. Since a proximinal space is necessarily closed, uniform boundedness of all irreducible lightning bolts in Q is also necessary for proximinality of D in $C(Q)$.

EXAMPLES 3.8. (a) The assumptions of Theorems 3.3 and 3.5 are satisfied for a wide class of sets. For example, they are satisfied for a curvilinear trapezoid

$$\{(x, y) : a \leq x \leq b,\ 0 \leq g \leq f(x)\}\,,$$

provided that $f(x)$ is a continuous function on $[a, b]$, or, also, the set

$$\{(x, y) : a \leq x \leq b,\ f_1(x) \leq y \leq f_2(x)\}$$

provided that $f_1'(x) f_2'(x) \leq 0$. The assumptions of Theorem 3.5 are also satisfied for a compact Q symmetric about the O_x-axis, provided that all its cross-sections by the lines $x = \text{const}$ are either segments or points. An arbitrary finite union of closed rectangular regions (with sides parallel to the coordinate axes) satisfies the assumptions of Theorem 3.5 as well. It is easy to form such a union that does not contain a global x-bar, y-bar, or, the more so, a cross.

(b) A closed region bounded by an ellipse whose axes are not parallel to the coordinate axes satisfies the assumptions of Theorem 3.3 but not those of Theorem 3.5. It is unknown whether D is proximinal there.

(c) Example 2.10d shows that the assumptions of Theorems 3.3 and 3.5 are essential (the assumptions of Theorem 3.5 are violated at the point $(\pi^2/6, \pi^2/6)$). In that example, the set Q is one-dimensional. Let us give an example showing that the assumptions of Theorems 3.3 and 3.5 are also essential when Q is a closed region. Let a closed region Q be bounded by the curves

$$y = x - \frac{1}{3}x^3, \quad y = x + \frac{1}{3}x^3, \quad x = 1.$$

The requirements of Theorem 3.5 fail either for points $x \in X = [0,1]$, or for points $y \in Y = \left[0, 1\frac{1}{3}\right]$. The requirements of Theorem 3.3 also fail: we can place in Q lightning bolts of an arbitrary length (Q/R consists of two classes: $(0,0)$ and $Q \setminus (0,0)$). Consider the functions

$$\varphi(x) = \begin{cases} \frac{1}{x}, & x \neq 0, \\ 0, & x = 0, \end{cases} \qquad \psi(y) = \begin{cases} -\frac{1}{y}, & y \neq 0, \\ 0, & y = 0, \end{cases}$$

and set $f(x,y) = \varphi(x)\psi(y)$. It is easy to check that $f \in C(Q)$, $f \in \overline{D}$, but $f \notin D$ and $f \notin \overline{BD}$; hence f neither has the best approximation in D, nor in BD. Yet $f \in \widetilde{BD}$, and hence is its own best approximation in $\widetilde{BD}$. We have $g(x,y) = \sin\varphi(x) + \sin\psi(y) \in C(Q)$, $g \in BD$, $g \in \overline{D}$, but $g \notin D$. So, if the requirements of Theorem 3.5 are not fulfilled, there exist situations in which a continuous function has a best approximation in BD, but not in D.

The results of these sections are taken from [**51**].

§4. Annihilator of sums of superpositions. When is the subspace of sums of superpositions everywhere dense?

In many problems of approximation by elements of a given subspace, it is necessary to make a study of the annihilator of that subspace. In this section we shall start the study of the annihilator of the subspace of sums of superpositions. First, consider a general case of linear superpositions. Let $X, X_1, \dots, X_i, \dots$ be compact sets, $\Phi_i : X \to X_i$ continuous mappings, and $h^i(x) \in C(X_i)$ given functions. Form the linear subspace in $C(X)$ that consists of all linear superpositions—each has its own number of terms:

$$D = \left\{ \sum_1^k h^i(x) g^i \circ \Phi_i(x),\ k \geq 1,\ g_i \in C(X_i) \right\}. \tag{4.1}$$

LEMMA 4.1. *In order that a measure $\mu \in C(X)^\star$ be orthogonal to the subspace* (4.1) *it is necessary and sufficient that*

$$\nu_i \overset{\text{def}}{=} \Phi_i \circ (h^i \mu) \equiv 0, \qquad i = 1, 2, \dots. \tag{4.2}$$

The proof follows immediately from formula (2.12) of Chapter 1. Let $\mu = \mu^+ - \mu^-$ be the Jordan decomposition of $\mu \in D^\perp$. Formula (4.2) carries some information concerning symmetry in the disposition of S_{μ^+} and S_{μ^-}. (S_ν is, as usual, the closed support of the measure ν.) Such symmetry appears more clearly when $h^i(x)=1$ for all $i=1,\dots$. In that case, (4.2) means that for an arbitrary Borel set $E\subset X_i$

$$\mu\left(\Phi_i^{-1}(E)\right)=0,$$

or

$$\mu^+\left(\Phi_i^{-1}(E)\right)=\mu^-\left(\Phi_i^{-1}(E)\right). \tag{4.3}$$

To make the condition (4.2) more transparent, consider the case when S_μ is a finite set.

COROLLARY 4.2. *Let $h^i(x)\equiv 1$, $i=1,\dots$, $\mu\in D^\perp$, and assume that the support S_μ of the measure μ is a finite set. Then each point $x_0\in S_{\mu^+}$ corresponds for all $i=1,2,\dots$ to at least one point $x_i\in S_{\mu^-}$ such that $\Phi_i(x_0)=\Phi_i(x_i)$ and, accordingly, each point $y_0\in S_{\mu^-}$ corresponds for all $i=1,\dots$ to at least one point $y_i\in S_{\mu^+}$ such that $\Phi_i(y_0)=\Phi_i(y_i)$.*

THEOREM 4.3 [**131**]. *Let X, $\{X_i)$ be compact metric spaces, and let D be defined by (4.1), where all $h^i(x)\equiv 1$, $i=1,2,\dots$. Let the operator $\tau\colon 2^X\to 2^X$ be defined by (7.1) and (7.2) of Chapter 1, where the intersection in the second formula is taken over all i. If $\mu\in D^\perp$ and $|\mu|$ denotes the total variation of μ, then*

$$|\mu|\left(X\backslash\bigcap_{n=1}^{\infty}\tau^n(X)\right)=0. \tag{4.4}$$

PROOF. As shown in Lemma 7.5 of Chapter 1, if $Z\subset X$ is a Borel set, then Z^i and $\tau(Z)$ are also Borel sets. For each i, consider $X\backslash X^i$. For any Borel set $E\subset X\backslash X^i$, we have

$$\mu(E)=\Phi_i\circ\mu\left(\Phi_i(E)\right)$$

and, in view of Lemma 4.1, $\mu(E)=0$. Hence, $|\mu|\left(X\backslash X^i\right)=0$. But then

$$|\mu|\left(X\backslash\tau(X)\right)=|\mu|\left(\bigcup_i\left(X-X^i\right)\right)\le\sum|\mu|\left(X\backslash X^i\right)=0,$$

so that the measure μ is supported on the set $\tau(X)$. Taking $Z=\tau(X)$, consider $Z\backslash\tau(Z)=\tau(X)\backslash\tau^2(X)$. Let $E\subset Z\backslash Z^i$. We have $E_1\overset{\text{def}}{=}\Phi_i^{-1}\left[\Phi_i(E)\right]=E\cup E_2$, where $E_2\subset X\backslash Z$. By Lemma 4.1, $\mu\left(E_1\right)=0$. By what we have already shown, $\mu\left(E_2\right)=0$, so $\mu(E)=0$. Thus, we have shown that $|\mu|\left(Z\backslash Z^i\right)=0$. Similarly, we can show now that $|\mu|\left(Z\backslash\tau(Z)\right)=0$ and, accordingly, the measure μ is supported on $\tau(Z)=\tau^2(X)$. Continuing these arguments, we can prove that for all n, $|\mu|\left(\tau^n(X)\backslash\tau^{n+1}(X)\right)=0$ and μ is supported on $\tau^n(X)$. Finally,

$$|\mu|\left(X\backslash\bigcap_{n=1}^{\infty}\tau^n(X)\right)=|\mu|\left(\bigcup_{n=1}^{\infty}\left(\tau^{n-1}(X)\backslash\tau^n(X)\right)\right)=0$$

and μ is supported on $\bigcap\limits_{n=1}^{\infty}\tau^n(X)$, while $S_\mu\subset\bigcap\limits_{n=1}^{\infty}\tau^n(X)$.

Approximation of measures in $D^\perp$ with finite supports. From now on, assume that D consists of functions (1.1) with a fixed number of terms N. In fact, without loss of generality, assume that $X \subset Y = X_1 \times \cdots \times X_N$, and the Φ_i are natural projections from Y onto X_i, $\Phi_i(X) = X_i$. We can consider functions in D as being defined on Y: $D = D(Y)$. We shall specifically say so whenever by Φ_i we understand its restriction to X (as, e.g., in (4.3)).

THEOREM 4.4. *Let $\mu \in D^\perp$, $S_\mu \subset X$, $\|\mu\| = 1$. There exists a set of measures $\{\mu_\alpha\} \subset D(Y)^\perp$ weak $(\star)$ converging in $C(Y)^\star$ to μ and satisfying the following properties*:

1. $\|\mu_\alpha\| = 1$.
2. *S_{μ_α} is a finite set.*
3. *For any open neighborhood $G \subset X$ there exists an index α_0 such that $S_{\mu_\alpha} \subset G$ for $\alpha > \alpha_0$.*

PROOF. To simplify notation, we shall conduct the arguments for $N = 2$. Consider partitions of X_1 into a finite number $U_1, \dots, U_k$ of Borel subsets, and similar partitions $V_1, \dots, V_\ell$ for X_2. For indices α we use all possible pairs of such partitions: $\alpha = (U_1, \dots, U_k; V_1, \dots, V_\ell)$. The set $\{\alpha\}$ has a natural partial order (according to the "fineness" of a partition), after which $\{\alpha\}$ becomes a directed set. Take some α. In each one of the sets $U_1, \dots, U_k$ that are incorporated into α, we choose a point $t_1, \dots, t_k$. Similarly, in $V_1, \dots, V_\ell$ we select points $\tau_1, \dots, \tau_\ell$. Points $x_{ij} = t_i \times \tau_j \in U_i \times V_j$, $i = 1, \dots, k$, $j = 1, \dots, \ell$, are points in $Y = X_1 \times X_2$. At each one of the points x_{ij} we place an atom of the measure $\tilde{\mu}_\alpha(x_{ij}) \stackrel{\text{def}}{=} \mu(U_i \times V_j)$. Applying Lemma 4.1 first to the measure μ, and then to a newly-constructed measure $\tilde{\mu}$, we conclude that $\tilde{\mu}_\alpha \in D(Y)^\perp$. Clearly, $\|\tilde{\mu}_\alpha\| \to 1$, so we can pass to measures $\mu_\alpha = \dfrac{\tilde{\mu}_\alpha}{\|\tilde{\mu}_\alpha\|}$ for which $\|\mu_\alpha\| = 1$.

The proof of (2) and (3) is a rather elementary exercise in measure theory, and we omit it. When X_1 and X_2 are compact metric spaces, instead of a net $\{\mu_\alpha\}$, one can construct an ordinary sequence $\{\mu_n\}$, and then the checking of properties (2) and (3) becomes especially simple.

Measures on lightning bolts. Consider the case $N = 2$ and return to the notation of (1.7) and (2.18) of the present chapter. If a lightning bolt $\ell = [p_1, \dots, p_n]$ contained in Q is finite, we associate with it a linear functional over $C(Q)$ defined by the formula

$$r_\ell(f) = \frac{1}{n}\sum_1^n (-1)^{i-1} f(p_i),$$

or

$$r_\ell(f) = \frac{1}{n}\sum_1^n (-1)^i f(p_i). \tag{4.5}$$

Thus, r_ℓ is generated by a measure that has atoms $\pm\dfrac{1}{n}$ with alternating signs at the vertices of the lightning bolt. We denote this measure by r_ℓ, the same as the corresponding functional, in order not to complicate the notation.

If δ_p is the delta-mass at p, then

$$r_\ell = \frac{1}{n}\sum_1^n (-1)^{i-1}\delta_{p_i}, \tag{4.6}$$

or

$$r_\ell = \frac{1}{n}\sum_1^n (-1)^{i}\delta_{p_i}. \tag{4.6$'$}$$

If $\ell : [p_1, \dots, p_n, \dots]$ is an infinite lightning bolt, we associate with it a sequence of linear functionals on $C(Q)$:

$$r_\ell^n = \frac{1}{n}\sum_1^n (-1)^{i-1} f(p_i), \qquad n = 1, 2, \dots. \tag{4.7}$$

LEMMA 4.5. 1. *For* (4.5) *we have* $\|r_\ell\| \le 1$, *while for* (4.7), $\|r_\ell^n\| \le 1$, $n = 1, 2, \dots$.

2. $\|r_\ell\| = 1$ $(\|r_\ell^n\| = 1, \quad n = 1, 2, \dots)$ *if and only if the set of vertices of the lightning bolt* ℓ *having even indices does not overlap with that having odd indices.*

3. *If* ℓ *is closed, then* $r_\ell \in D^\perp$. *If* ℓ *is an infinite lightning bolt and* r *is a limit point of the sequence of linear functionals* $\{r_\ell^n\}$ *in the weak* $(\star)$ *topology of the space* $C(Q)^\star$, *then* $r \in D^\perp$.

PROOF. 1. Clearly, $|r_\ell(f)| \le \frac{1}{n} n\|f\|$, and hence $\|r_\ell\| \le 1$. Similarly, $\|r_\ell^n(f)\| \le 1$, $n = 1, 2, \dots$.

2. The assumptions in part 2 of the lemma mean that the same point p cannot simultaneously be a vertex of ℓ having even and odd indices (although we do not exclude cases when among the vertices of ℓ having even indices (or odd, for that matter) there are equal ones). Consider the case of a finite lightning bolt $\ell = [p_1, \dots, p_n]$. In the case when a vertex p is repeated among vertices with odd indices m times it accumulates the charge $\frac{m}{n}$. If a vertex p is repeated k times among vertices with even indices, it accumulates the charge $-\frac{k}{n}$. Clearly, in this situation we always have

$$\int_Q |dr_\ell| = \int_\ell |dr_\ell| = \frac{1}{n} \cdot n = 1,$$

i.e., $\|r_\ell\| = 1$, and, similarly, $\|r_\ell^n\| = 1$, $n = 1, 2, \dots$, for the case of an infinite bolt. If a vertex p had both even and odd indices, then in the sum (4.6) (or (4.7)) the corresponding terms would cancel, since they have opposite signs. Thus, the total variation of the charge will still be less than $1 - \frac{2}{n}$.

3. Let, e.g., $\ell = p_1 R_1 p_2 R_2 p_3 \dots p_{n-1} R_{n-1} p_n$. If n is even, then $R_{n-1} = R_1$, while if it is odd, then $R_{n-1} = R_2$. In the first case ($n = 2m$) for an arbitrary sum $g_1 \circ \pi_1(p) + g_2 \circ \pi_2(p)$ we have

$$r_\ell\left[g_1 \circ \pi_1 + g_2 \circ \pi_2\right] = \frac{1}{n}\left[g_2\left(\pi_2\left(p_1\right)\right) - g_2\left(\pi_2\left(p_n\right)\right)\right]. \tag{4.8}$$

In the second case ($n = 2m+1$) we have

$$r_\ell \left[g_1 \circ \pi_1 + g_2 \circ \pi_2\right] = \frac{1}{n}\left[g_2\left(\pi_2(p)\right) + g_1\left(\pi_1\left(p_n\right)\right)\right]. \tag{4.8'}$$

Similar formulas also hold when $p_1 R_2 p_2$. If the lightning bolt is closed, then $n = 2m$ and (4.8) implies that $r_\ell \in D(Q)^\perp$. If ℓ is infinite and a subsequence $\{r_\ell^{n_k}\}$, $n_k \to \infty$, converges to a functional r in the weak ($\star$) topology of $C(Q)^\star$, then it follows from (4.8) and (4.8′) that $r \in D^\perp$. (Note that we do not rule out the case $r \equiv 0$.)

Below, we shall mainly use only the lightning bolts for which $\|r_\ell\| = 1$ (and therefore, the sets of vertices with even or odd indices are disjoint). This is stipulated by the fact that in duality relations one uses measures $\mu \in D^\perp$ for which $\|\mu\| = 1$.

Representation of measures in $D^\perp$ with finite supports by measures on lightning bolts.

LEMMA 4.6. *If $\mu \in D^\perp$, $\|\mu\| = 1$ and μ has support S_μ that consists of finitely many points, then μ is a convex combination of measures on closed lightning bolts whose supports are contained in S_μ.*

PROOF (V. A. Medvedev). Choose an arbitrary point $p_1 \in S_\mu$. According to (2.2) and (2.3) there exists a point $p_2 \in S_\mu$, $p_1 R_1 p_2$, while $\mu(p_2)\mu(p_1) < 0$. Further, there exists $p_3 \in S_\mu$ with $p_2 R_2 p_3$ and $\mu(p_2)\mu(p_3) < 0$, etc. The process is finished when at the next step we arrive at one of the points constructed earlier. Without loss of generality assume that p_1 is such a point. Let us assume that the constructed points $p_1, p_2, \dots, p_n$, p_1 form a circular lightning bolt. If $p_n R_2 p_1$, then the lightning bolt $\ell_1 = [p_2, \dots, p_n]$ is closed. At the neighboring vertices of ℓ_1 atoms of μ have opposite signs. Let m be the smallest of the absolute values of atoms of the measure μ at vertices of ℓ_1. Consider measure r_{ℓ_1}, choosing the signs of the atoms on ℓ_1 to be the same as those for μ. If N denotes the number of vertices of ℓ_1 ($N = n$ or $N = n-1$), then the measure Nmr_{ℓ_1} has values $\pm m$ at the vertices of ℓ_1 and the signs coincide with those of μ. Since m is the smallest modulus of values of μ at the vertices of ℓ_1, the measure $\mu_1 = \mu - \lambda_1 r_{\ell_1}$, where $\lambda_1 = Nm$, has at the vertices of ℓ_1 the same signs as μ, and $|\mu_1| = |\mu| - Nm\,|r_{\ell_1}|$; hence $\|\mu_1\| = \|\mu\| - \lambda_1\|r_{\ell_1}\| = 1 - \lambda_1$. Further, $r_{\ell_1} \in D^\perp$. So $\mu_1 \in D^\perp$. Finally, at a vertex of ℓ_1 where μ takes a value $\pm m$, the value of μ_1 equals zero. So, the support of μ_1 is smaller than S_μ by at least one point. Now repeat a similar construction on S_{μ_1}. Clearly, after finitely many steps we arrive at

$$\mu = \sum_{k=1}^{\nu} \lambda_k r_{\ell_k}, \quad \sum_1^{\nu} \lambda_k = 1, \qquad \lambda_k > 0.$$

COROLLARY 4.7. *For $N = 2$, in view of Theorem 4.4, each one of the measures μ_α approximating the measure $\mu \in D(Q)^\perp$ is a convex combination of measures on closed lightning bolts situated in G for $\alpha > \alpha_0$.*

Duality relations. Let L be a normed linear space, L_0 a subspace in L and f an element of L. The following duality relation has by now become a routine part of approximation theory:

$$\max_{\substack{\ell \in L_0^\perp \\ \|\ell\| \leq 1}} |\ell(f)| = \inf_{g \in L_0} \|f - g\|. \tag{4.9}$$

(Here, as always, we write max (min) instead of sup (inf) when we can guarantee that it is attained.) Using the earlier notation, let us consider a problem of the best approximation in $C(Q)$ of a given function $f \in C(Q)$ by functions in the subspace D:

$$E(f) \overset{\text{def}}{=} \inf_{g \in D} \|f - g\| = \inf_{\substack{g_1 \in C(X) \\ g_2 \in C(Y)}} \max_{p \in Q} |f(p) - (g_1 \circ \pi_1(p) + g_2 \circ \pi_2(p))| . \tag{4.10}$$

Then, (4.9) becomes

$$E(f) = \max_{\substack{\mu \in D^\perp \\ \|\mu\| \leq 1}} \left| \int_Q f d\mu \right| . \tag{4.11}$$

Theorem 4.4 and Corollary 4.7 allow us to use in (4.11) measures with a simple structure, although without being able to guarantee that sup is attained. Assume that all functions in $C(Q)$ are extended to functions in $C(X \times Y)$, and all functions in $D(Q)$ to functions in $D(X \times Y)$.

Let G be an open neighborhood of Q, and let $\eta = G$ run over the directed set of all such neighborhoods of Q partially ordered in the usual way. Set

$$R_\eta(f) = \sup_{\ell \subset G} \left| \int_\ell f dr_\ell \right| = \sup_{\ell \subset G} |r_\ell(f)| , \tag{4.12}$$

where sup is taken over all closed lightning bolts ℓ. Clearly, the scalar-valued net $\{R_\eta(f)\}$ is decreasing with respect to η. Therefore, there exists

$$\lim_\eta R_\eta(f) = \inf_{G \supset Q} R_\eta(f). \tag{4.13}$$

THEOREM 4.8 (duality). *If $Q = X \times Y$, then*

$$E(f) = \sup_{\ell \in Q} |r_\ell(f)| , \tag{4.14}$$

where the supremum is taken over all closed lightning bolts ℓ. In the general case $Q \subset X \times Y$,

$$E(f) = \lim_\eta R_\eta(f). \tag{4.15}$$

PROOF. Fix $\varepsilon > 0$, and let $g_1 \in C(X)$ and $g_2 \in C(Y)$ be such that

$$\|f - g_1 \circ \pi_1(p) - g_2 \circ \pi_2(p)\|_Q < E(f) + \varepsilon.$$

Take a domain $\eta = G \subset Q$ so that

$$\|f - g_1 \circ \pi_1(p) - g_2 \circ \pi_2(p)\|_{\overline{G}} < E(f) + 2\varepsilon.$$

For any closed lightning bolt $\ell \subset G$ we have, since $r_\ell \in D^\perp$,

$$\begin{aligned} |r_\ell(f)| &= |r_\ell[f - g_1 \circ \pi_1(p) - g_2 \circ \pi_2(p)]| \\ &\leq \|f - g_1 \circ \pi_1 - g_2 \circ \pi_2\| < E(f) + 2\varepsilon. \end{aligned} \tag{4.16}$$

Therefore,

$$R_\eta(f) < E(f) + 2\varepsilon.$$

So, we obtain that

$$\lim_\eta R_\eta(f) \leq E(f). \tag{4.17}$$

On the other hand, let $\mu^\star \in D^\perp(Q)$ be a measure that realizes the max in (4.11). Then $\|\mu^\star\| = 1$ and $S_{\mu^\star} \subset Q$. According to Theorem 4.4 and Corollary 4.7, for an arbitrary open set $G \supset Q$ there exists a measure μ such that $S_\mu \subset G$, $\|\mu\| = 1$, and

$$\mu = \sum_{\theta=1}^{s'} \lambda_\theta r_{\ell_\theta}, \qquad \sum_1^s \lambda_\theta = 1, \qquad \lambda_\theta > 0, \tag{4.18}$$

where the ℓ_θ are closed lightning bolts in G and

$$\left| \int_G f\,d\mu - \int_Q f\,d\mu^\star \right| < \varepsilon. \tag{4.19}$$

From (2.34) we obtain that

$$\left| \int_G f\,d\mu \right| > \left| \int_Q f\,d\mu^\star \right| - \varepsilon = E(f) - \varepsilon. \tag{4.20}$$

Since μ is a convex combination of the r_{ℓ_θ}, there exists an index θ such that

$$\left| \int_G f\,dr_{\ell_\theta} \right| = |r_{\ell_\theta}(f)| \geq \left| \int_G f\,d\mu \right| > E(f) - \varepsilon. \tag{4.21}$$

Hence ($\eta = G$),

$$R_\eta(f) > E(f) - \varepsilon \quad \text{and} \quad \lim_\eta R_\eta(f) \geq E(f). \tag{4.22}$$

Combining (4.22) and (4.17), we complete the proof of the theorem.

COROLLARY 4.9. *If $E(f) > 0$, then among the measures realizing the* max *in* (4.11) *there exists a measure equal to the weak* ($\star$) *limit of a net of measures* $\{r_{\ell_\alpha}\}$ *supported on closed lightning bolts $\ell_\alpha \subset X \times Y$, and the net $\{\ell_\alpha\}$ is such that for any open net $G \supset Q$ there exists an index α_0 for which $\ell_\alpha \subset G$ for all $\alpha > \alpha_0$.*

The proof is omitted, since it is similar to that of Corollary 4.12 below.

Bibliographical notes. The duality theorem in the case when Q is a rectangle in $\mathbb{R}^2$ was proven in the seminal paper of Diliberto and Straus [**38**]. At that time, the duality approach had not yet become widely popular in approximation theory. (Papers by Krein and Nikol′skiĭ [**120**], [**89**] played a crucial role in spreading duality methods to problems of best approximation.) The theorem was independently found by S. A. Smolyak, and some ideas of the proof were suggested by Arnold (cf. [**113**]). Yet another proof was given in Golomb's paper [**64**], another of the few very first publications concerning these problems. However, the latter proof had an essential gap, pointed out in [**100**]. For Q's other than a rectangle, the theorem appeared in [**80**]. In [**79**] another form of the result was given under the additional assumption that the best approximation in D does exist. In [**100**] the theorem was proved without this additional assumption. We shall discuss some of these results later on.

When is $D(Q)$ dense in $C(Q)$?

COROLLARY 4.10. *If a subspace D is as in (4.1), in order that D be everywhere dense in $C(X)$ it is necessary and sufficient that (4.2) imply $\mu \equiv 0$.*

COROLLARY 4.11. *In order that a subspace in Theorem 4.4 be everywhere dense in $C(X)$, it is necessary and sufficient that any net of measures $\{\mu_\alpha\}$ satisfying all the properties listed in Theorem 4.4 converge weak $(\star)$ to zero in $C(Y)^\star$.*

COROLLARY 4.12. *For $N = 2$, under the assumptions (and notation) of Corollary 4.7 the following conditions are equivalent:*
1. $\overline{D(Q)} = C(Q)$.
2. *Let $\{r_{\ell_\alpha}\}$ be a net of measures on closed lightning bolts $\{\ell_\alpha\}$ such that for an arbitrary open set $G \supset Q$ there exists an index α_0 such that $\ell_\alpha \subset G$ for $\alpha > \alpha_0$. Then, $\{r_{\ell_\alpha}\}$ converges weak $(\star)$ to zero in $C(X \times Y)^\star$.*

We only need to prove Corollary 4.12. Assume that $\overline{D} \neq C(Q)$. Then there exists $f \in C(Q)$ such that $d = \operatorname{dist}(f, D) > 0$. According to the Krein–Nikol′skiĭ duality relation (4.11), there exists a measure $\mu \in D(Q)^\perp$, $\|\mu\| = 1$, such that

$$\int_Q f d\mu = d.$$

Let $\{\mu_\alpha\}$ be a net of measures in $C(X \times Y)^\star$ approximating μ in accordance with Theorem 4.4. By Corollary 4.7 each one of the measures μ_α for $\alpha > \alpha_0$ is a convex combination of measures on closed lightning bolts located in G. For at least one of those bolts ℓ_α

$$\int f dr_{\ell_\alpha} \geq \int f d\mu_\alpha,$$

and hence $\{r_{\ell_\alpha}\}$ cannot converge to zero.

From Theorem 4.3 we obtain

COROLLARY 4.13. *If under the assumptions of Theorem 4.3 we have*

$$\bigcap_{n=1}^{\infty} \tau^n(X) = 0,$$

then $\overline{D} = C(X)$.

A geometric condition for density of $D(Q)$. Let us clarify the "geometric" meaning of Corollary 4.12. Let ℓ be a finite lightning bolt and g an open set. Denote by $S(\ell)$ the number of vertices of ℓ and by $S^+(\ell, g)$ $(S^-(\ell, g))$ the number of vertices of ℓ inside g with positive (negative) masses.

PROPOSITION 4.14. *Let $\{\ell_\alpha\}$ be a net of lightning bolts. If for any open set g we have*

$$\left|S^+(\ell_\alpha, g) - S^-(\ell_\alpha, g)\right| = o\left(S\left(\ell_\alpha\right)\right), \tag{4.23}$$

then the set $\{r_{\ell_\alpha}\}$ weak $(\star)$ converges to zero in $C(X \times Y)^\star$.

PROOF. (4.23) can be rewritten as

$$r_{\ell_\alpha}(g) = o(1). \tag{4.24}$$

Then use a general result of Alexandrov on weak ($\star$) convergence (Theorem IV.9.15 in [**40**]). The fact that there the theorem is stated for sequences, while here we are dealing with nets, is of no importance.

The Marshall–O'Farrell criterion. For $N = 2$ there exists a much more effective criterion for $D = D(Q)$ to be everywhere dense in $C(Q)$, yet with an essential additional assumption.

THEOREM 4.15. *In order that $\overline{D} = C(Q)$, it is necessary and, if for each point $P \in Q$ the equivalence class $R(P)$ is closed, it is also sufficient that there exist no closed lightning bolts in Q.*

LEMMA 4.16. *Let $E = R(E_1)$, where E_1 is a Borel set. If a measure $\mu \in D(Q)^\perp$, then the restriction $\nu = \mu|_E$ of μ on E also belongs to $D(Q)^\perp$.*

PROOF. Since $\mu \in D(Q)^\perp$, according to Lemma 4.1 we have $\mu_i = \pi_i \circ \mu = 0$, $i = 1, 2$. Set $\nu_i = \pi_i \circ \nu$. We must show (again, in view of Lemma 4.1) that $\nu_i = 0$. Let A be a Borel subset in X, and let $A_1 = \pi_1(E)$. Since E is saturated with respect to R, then whenever $p \in E$, $\pi_1^{-1}\pi_1(p) \subset E$ as well. Therefore,

$$\begin{aligned}\nu_1(A) =& \nu_1(A \cap A_1) + \nu_1(A \backslash A \cap A_1) = \nu\left[\pi_1^{-1}(A \cap A_1)\right] + \nu\left[\pi_1^{-1}(A \backslash A \cap A_1)\right]\\ =& \mu\left[\pi_1^{-1}(A \cap A_1)\right] + \nu\left[\pi_1^{-1}(A \backslash A \cap A_1)\right]\\ =& \mu\left[\pi_1^{-1}(A \cap A_1)\right] + 0 = \mu_1(A \cap A_1) = 0.\end{aligned}$$

Hence $\nu_1 = 0$. Similarly, one can show that $\nu_2 = 0$.

Extreme points in $D(Q)^\perp$. If $D^\perp = \{0\}$, $\overline{D} = C(Q)$. Assuming then that $D(Q)^\perp \neq \{0\}$, consider a set σ of regular signed Borel measures μ on Q defined by

$$\sigma = \left\{\mu : \mu \in D(Q)^\perp,\ \|\mu\| \leq 1\right\}. \tag{4.25}$$

σ is convex and closed in the weak ($\star$) topology of the space $C(Q)^\star$, and so, according to the Krein–Milman theorem (cf., e.g., [**40**, Chapter V, Section 8]), has extreme points and is equal to the weak ($\star$) closure of convex combinations of its extreme points.

LEMMA 4.17. *Assume that the equivalence class $R(p)$ of any point p is a closed set. If $F_1 \supset F_2 \supset \cdots \supset F_n \supset \ldots$ is a sequence of closed sets in a compact Q, then*

$$\bigcap_{n=1}^{\infty} R(F_n) = R\left(\bigcap_{n=1}^{\infty} F_n\right). \tag{4.26}$$

PROOF. Let $p \in \bigcap_{n=1}^{\infty} R(F_n)$. Then, for all n there exists $p_n \in F_n$ such that $p = R(p_n)$ and $p_n = R(P)$. Since Q is compact, the sequence $\{p_n\}$ has a limit point p_0. In view of the closedness of all F_n's, $p_0 \in \bigcap_{n=1}^{\infty} F_n$. Since $R(p)$ is a closed set too,

$p_0 \in R(p)$, $p \in R(p_0)$. Thus, $p \in R\left(\bigcap_{n=1}^{\infty} F_n\right)$. Hence, $\bigcap_{n=1}^{\infty} R(F_n) \subseteq R\left(\bigcap_{n=1}^{\infty} F_n\right)$. The opposite inclusion is trivial.

LEMMA 4.18. *Suppose that for all points $p \in Q$ the equivalence classes $R(p)$ are closed. If μ is an extreme point of the unit ball σ (see (4.25)), then the closed support S_μ of the measure μ is contained in some equivalence class.*

PROOF. For any set E let

$$\gamma(E) = R_2\left(R_1(E)\right] = \pi_2^{-1} \circ \pi_2 \circ \pi_1^{-1} \circ \pi_1(E). \tag{4.27}$$

Then, $\gamma^n(E)$ is the set of all points that can be joined to points in E by lightning bolts of length $\leq 2n$. Let $p \in S_\mu$, while $q \notin R(p)$. Choose a neighborhood of the point q whose closure U does not intersect $R(p)$. Since $p \notin \gamma^n(U)$ for all n and U is closed, $\gamma^n(U)$ is also closed and there exists a closed neighborhood V_n of the point p that does not intersect $\gamma^n(U)$. Without loss of generality assume that $V_1 \supset V_2 \supset \dots \supset V_n \supset \dots$ We have

$$R(U) = \bigcup_{n=1}^{\infty} \gamma^n(U) \quad \text{and} \quad V_n \cap \gamma^n(U) = \emptyset.$$

Moreover,

$$\begin{aligned}\left(\bigcap_{m=1}^{\infty} V_m\right) \cap R(U) = \bigcup_{n=1}^{\infty} \left(\bigcap_{m=1}^{\infty} V_m\right) \cap \gamma^n(U) = \emptyset \\ \Leftrightarrow U \cap R\left(\bigcap_{m=1}^{\infty} V_m\right) = \emptyset \Leftrightarrow U \cap \bigcap_{m=1}^{\infty} R(V_m) = \emptyset.\end{aligned}$$

Since $p \in S_\mu$ and V_m is a neighborhood of p, we must have $|\mu|(V_m) > 0$. Then, however, the more so $|\mu|\left(R(V_m)\right) = \alpha > 0$. Let us show that $\alpha = 1$. If $\alpha < 1$, set $\mu_1 = \mu\big|_{R(V_m)}$ and $\mu_2 = \mu\big|_{Q\setminus R(V_m)}$. In view of Lemma 2.17, $\mu_1 \in D(Q)^\perp$ and $\mu_2 \in D(Q)^\perp$, while $\|\mu_1\| = \alpha$, $\|\mu_2\| = 1-\alpha$, and $\mu = \alpha\frac{\mu_1}{\alpha} + (1-\alpha)\frac{\mu_2}{1-\alpha}$. Therefore, μ cannot be an extreme point. Hence, $\alpha = |\mu|\left(R(V_m)\right) = 1$. Since $\{R(V_m)\}$ is a decreasing sequence of sets, we have

$$|\mu|\left(\bigcap_{m=1}^{\infty} R(V_m)\right) = \lim_{m\to\infty} |\mu|\left(R(V_m)\right) = 1.$$

Hence, $|\mu|(U) = 0$ and $q \notin S_\mu$.

Note that the contents of Lemma 4.18 is given in [**99**] by one sentence without any hint regarding its proof. S. Ya. Khavinson has reproduced Lemma 4.18 for compact metric spaces. The above presentation (together with Lemma 4.17) was suggested by V. A. Medvedev.

LEMMA 4.19. *Under the assumptions of Lemma* 4.18 *the equivalence class $R(p)$ that contains S_μ must contain a closed lightning bolt.*

PROOF. For the sake of brevity, set $A = R(p)$ and let $A_1 = \{p\}$, $A_2 = \pi_1^{-1} \circ \pi_1(A_1)$, $A_3 = \pi_2^{-1} \circ \pi_2(A_2)$, ... ,

$$A_n = \pi_j^{-1} \circ \pi_j(A_{n-1}), \qquad j = 1, 2, \tag{4.28}$$

where j and $n - 1$ are simultaneously even or odd. All A_n are compact, and

$$R(p) = A = \bigcup_{n=1}^{\infty} A_n. \tag{4.29}$$

Since $A_n \subseteq A_{n+1}$, we have

$$\lim_{n\to\infty} \mu^+(A_n) = \lim_{n\to\infty} \mu^-(A_n) = \frac{1}{2},$$

where $\mu = \mu^+ - \mu^-$ ($|\mu| = \mu^+ + \mu^-$) is the Jordan decomposition of the measure μ. (Recall that $\mu(A) = 0$ and $|\mu|(A) = 1$.) Let n_0 be such that $\mu^+(A_{n_0}) > \frac{1}{3}$. According to the Hahn decomposition of the set A_{n_0} there exists a Borel set $E_0 \subset A_{n_0}$ such that

$$\mu^+(E_0) > \frac{1}{3} \quad \text{and} \quad \mu^-(E_0) = 0.$$

Without loss of generality assume n_0 to be even. In view of Lemma 4.1,

$$\mu\left[\pi_2^{-1} \circ \pi_2(E_0)\right] = 0.$$

Since $E_0 \subset \pi_2^{-1} \circ \pi_2(E_0)$, $\mu^+\left[\pi_2^{-1} \circ \pi_2(E_0)\right] = \mu^-\left[\pi_2^{-1} \circ \pi_2(E_0)\right] > \frac{1}{3}$. According to Hahn's theorem, there exists a Borel set $E_1 \subset \pi_2^{-1} \circ \pi_2(E_0) \subset A_{n_0+1}$ such that

$$\mu^-(E_1) > \frac{1}{3} \quad \text{and} \quad \mu^+(E_1) = 0, \qquad E_0 \cap E_1 = \emptyset.$$

Continue this process by considering the set $\pi_1^{-1} \circ \pi_1(E_1) \subset A_{n_0+2}$. We arrive at the set $E_2 \subset A_{n_0+2}$ with the properties:

$$\mu^+(E_2) > \frac{1}{3}, \qquad \mu^-(E_2) = 0, \qquad E_1 \cap E_2 = \emptyset,$$

and so on. As a result, we obtain a sequence of Borel sets $E_0, E_1, E_2, \ldots$ with the following properties:

$$\begin{cases} E_i \subset A_{n_0+i}, & i = 0, \ldots; \\ E_{i+1} \subset \pi_j^{-1} \circ \pi_j(E_i), & i = 0, \ldots, \quad j - 1, 2; \\ & i, j \text{ simultaneously even or odd;} \\ \mu^+(E_{2s}) > \frac{1}{3} \text{ and } \mu^-(E_{2s}) = 0, & s = 0, 1, \ldots; \\ \mu^-(E_{2s+1}) > \frac{1}{3} \text{ and } \mu^+(E_{2s+1}) = 0, & s = 0, 1, \ldots; \\ E_i \cap E_k = \emptyset & \text{whenever } i \text{ is even and} \\ & k \text{ odd, or vice versa.} \end{cases} \tag{4.30}$$

The latter condition follows from the fact that all the E_{2s} are taken inside the support of μ^+ while all the E_{2s+1} are taken from inside the support of μ^-, and those supports do not overlap by Hahn's theorem. On the other hand, any two sets E_i and E_j with $i, j = 2s$ must intersect. Indeed, assuming that $E_i \cap E_j = \emptyset$, and adding another set E_k with k having parity difference from that of i and j, we obtain three pairwise non-overlapping sets E_i, E_j, E_k, and then we have

$$1=|\mu|(Q)\geq|\mu|(E_i)+|\mu|(E_j)+|\mu|(E_k)>3\cdot\frac{1}{3}=1,$$

a contradiction. Take a set E_{2s}, $s>n_0$. Then, $E_0\cap E_{2s}\neq\emptyset$. Let $q\in E_0\cap E_{2s}$. According to the construction and properties of sets E_n there exists a lightning bolt

$$(4.31)\qquad [q_0,q_1,\dots,q_{2s-1},q],\qquad q_i\in E_i,\quad i=0,1,\dots,2s-1.$$

If among the vertices of this bolt there are repeating ones, then (4.31) contains a circular lightning bolt, and therefore it does contain a closed one and the lemma is proved. Thus, we assume all the vertices in (4.31) to be distinct. Since $q\in E_0$ and $q_0\in E_0$ while $E_0\subseteq A_{n_0}$, according to the construction of the sets A_i there exist lightning bolts joining p and q, p and q_0:

$$(4.32)\qquad [p,p_1,\dots,p_r,q_0]\quad\text{and}\quad [p,p_1',\dots,p_t',q],$$

where $[p_1,\dots,p_r]$ and $[p_1',\dots,p_t']$ in any case belong to A_{n_0} and $r<n_0$, $t<n_0$. Consider the sequence

$$(4.33)\qquad p,p_1,\dots,p_r,q_0,q_1,\dots,q_{2s-1},q,p_t',\dots,p_1',p.$$

The sequence (4.33) need not be a lightning bolt due to the behavior at q and q_0 of the junction of (4.31) with (4.32). However, performing an obvious reconstruction of the links at those points and some of the following points, we obtain a circular lightning bolt. Here the main point is that the inequality $s>n_0$ and the fact that all vertices in (4.31) are distinct guarantee that we do not return from q to p over the same sequence of vertices used to reach q from p. From a circular lightning bolt we choose a closed one, and the lemma is proved.

PROOF OF THEOREM 4.15. *Necessity.* Let Q contain a closed lightning bolt ℓ. By Lemma 4.5 the linear functional r_ℓ (see (4.7)) belongs to $D^\perp$. Therefore, $D(Q)$ is not dense in $C(Q)$.

Sufficiency. Consider $D(Q)^\perp$. If $D^\perp\neq\emptyset$, then according to Lemmas 4.18 and 4.19 we could find a closed lightning bolt in Q.

Extreme points in $D(Q)^\perp$ among measures with finite support. Lemma 4.6 shows that among measures with finite support only measures corresponding to closed lightning bolts can serve as extreme points in the unit ball σ in $D(Q)^\perp$. Let us clarify more precisely the conditions that a closed lightning bolt ℓ must satisfy in order that r_ℓ be an extreme point in σ.

LEMMA 4.20. *Let there be given measures $\mu_i\in C(Q)^\star$, $i=0,1,\dots,k$, and let*

$$(4.33')\qquad \mu_0=t_1\mu_1+\dots+t_k\mu_k,\qquad t_i>0,\quad\sum_1^k t_i=1$$

and $\|\mu_i\| = 1$, $i = -, 1, \dots, k$. *Then,*

$$\begin{gathered}\bigcup_{i=1}^{k} S_{\mu_i} \subseteq S_{\mu_0}, \qquad \bigcup_{i=1}^{k} Q_i^+ \subset Q_0^+, \qquad \bigcup_{i=1}^{k} Q_i^- \subset Q_0^-, \\ \mu_0^+ = \sum_1^k t_i \mu_i^+, \qquad \mu_0^- = \sum_1^n t_i \mu_i^-,\end{gathered} \tag{4.34}$$

where $Q_i^+ \cup Q_i^-$ *is the Hahn decomposition for* μ_i, $\mu_i = \mu_i^+ - \mu_i^-$, $|\mu_i| = \mu_i^+ + \mu_i^-$ *is the Jordan decomposition for* μ_i, *and* S_{μ_i} *is the closed support of the measure* μ_i.

PROOF. All relations (4.34) are almost obvious corollaries of (4.33′) and the fact that the total variations of all the measures are the same. For instance, we have

(4.35)

$$\mu_0\left(Q_0^+\right) = \mu_0^+\left(Q_0^+\right) = \sum_{i=1}^{k} t_i \mu_i\left(Q_0^+\right) \le \sum_{i=1}^{k} t_i \left|\mu_i\left(Q_0^+\right)\right| \le \sum_{i=1}^{k} t_i \|\mu_i\|_{Q_0^+},$$

(4.35′)

$$\mu_0^-\left(Q_0^-\right) = -\mu_0\left(Q_0^-\right) = -\sum_1^k t_i \mu_i\left(Q_0^-\right) \le \sum_1^k t_i \left|\mu_i\left(Q_0^-\right)\right| \le \sum_1^k t_i \|\mu_i\|_{Q_0^-}.$$

All the inequalities (4.35)–(4.35′) are in fact equalities, and we obtain the inclusions and equalities (4.34).

Let $\ell = [p_1, \dots, p_{2n}]$ be a closed lightning bolt. A closed lightning bolt $L = [q_1, \dots, q_{2m}]$, $m < n$, is called an oriented regular part of ℓ if

1. all vertices of L are contained among the vertices of ℓ, and
2. the signs of charges at the vertices $q_1, \dots, q_{2n}$ of ℓ alternate in the same way as in L.

EXAMPLE 4.21. Consider (in R^2) a closed lightning bolt $\ell = [p_1, \dots, p_{12}]$, where $p_1(4,0)$, $p_2(4,2)$, $p_3(3,2)$, $p_4(3.3)$, $p_5(4,3)$, $p_6(4,4)$, $p_7(1,4)$, $p_8(1,3)$, $p_9(0,3)$, $p_{10}(0,2)$, $p_{11}(1,2)$, $p_{12}(1,0)$. Agree, when writing an oriented lightning bolt, to begin with a vertex having a positive charge. The lightning bolts $L_1 = [p_1, p_2, p_{11}, p_{12}]$; $L_2 = [p_{11}, p_{10}, p_9, p_8]$; $L_3 = [p_3, p_2, p_5, p_4]$; $L_4 = [p_1, p_6, p_7, p_{12}]$; $L_5 = [p_3, p_4, p_9, p_{10}]$; $L_6 = [p_5, p_6, p_7, p_8]$ all give examples of oriented regular parts of ℓ. The closed lightning bolt $L_7 = [p_3, p_4, p_8, p_{11}]$ is not an oriented regular part of ℓ.

PROPOSITION 4.22. 1. *Let* ℓ *be a closed lightning bolt. The measure* r_ℓ *corresponding to* ℓ *is an extreme point of the unit ball in* $D(Q)^\perp$ *if and only if* ℓ *does not contain oriented regular parts.*

2. *If*

$$r_\ell = \sum_1^k t_i r_{\ell_i}, \qquad t_i > 0, \quad \sum_1^k t_i = 1, \tag{4.36}$$

then each one of the closed lightning bolts ℓ_i *is an oriented regular part of* ℓ. *Representation* (4.36) *is not unique in general.*

PROOF. The second statement follows directly from Lemma 4.20. We prove the first statement. If r_ℓ is not an extreme point, then we have $r_\ell = \frac{1}{2}[\mu_1 + \mu_2]$, where $m_i \in D(Q)^\perp$, $\|\mu_i\| = 1$, $i = 1, 2$, and $\mu_1 \neq \mu_2$. In view of Lemma 4.6 we obtain a representation of r_ℓ in the form (4.36). But then, according to the second statement, ℓ has oriented regular parts. Now, conversely, let ℓ have a regular oriented part L. If the number of vertices in ℓ equals $2n$ while in L it equals $2m$, then $\mu = r_\ell - \dfrac{m}{n} r_L \in D(Q)^\perp$. μ vanishes at those vertices of ℓ that belong to L, and is the same as for ℓ elsewhere. Therefore, $\|\mu\| = 1 - \dfrac{m}{n}$. The measure μ has a finite support, and hence, according to Lemma 4.6, μ can be represented by (4.36). So, the measure r_i is also given by a convex combination (4.36). In Example 4.21 we have, e.g.,

$$r_\ell = \frac{1}{3}\left(r_{L_1} + r_{L_5} + r_{L_6}\right) = \frac{1}{3}\left(r_{L_2} + r_{L_3} + r_{L_4}\right).$$

EXAMPLE 4.23. Any lightning bolt ℓ with four vertices has r_ℓ as an extreme point. Another example: the lightning bolt $p_1(0,0)$, $p_2(1,0)$, $p_3(1,1)$, $p_4(1/2,1)$, $p_5(1/2,2)$, $p_6(0,2)$. Proposition 4.22 is due to S. Ya. Khavinson and V. A. Medvedev.

Further information concerning extreme points of the unit ball in $D(Q)^\perp$, densities in $D(Q)$ and duality. Let us list without proofs some additional information concerning the extreme points in $D(Q)^\perp$. Let $\ell = [p_1, p_2, \dots]$ be an infinite lightning bolt and let r_ℓ^n, $n = 1, 2, \dots$, be signed measures defined by (4.7). If a measure μ is equal to the limit of the sequence $\{r_\ell^n\}$ in the weak $(\star)$ topology of $C(Q)^\star$, we shall say that μ is induced by ℓ. If ℓ is a finite closed lightning bolt, then the corresponding measure r_ℓ can be viewed as that induced by an infinite lightning bolt obtained from ℓ by circling it infinite many times. In [**100**], Marshall and O'Farrell obtained in principle a complete description of extreme points in $D(Q)^\perp$.

THEOREM 4.24 (Marshall–O'Farrell). *Let Q be a compact metric space. If a measure μ is an extreme point of the unit ball in $D(Q)^\perp$, then it is induced by an infinite lightning bolt ℓ.*

This result (in fact, [**100**] has it in a much more complete form) is obtained by using essentially new techniques from ergodic theory and stochastic processes.

COROLLARY 4.25 [**100**]. *In order that $D(Q)$ (Q is a compact metric space) be everywhere dense in $C(Q)$, it is necessary and sufficient that for any infinite lightning bolt ℓ the sequence $\{r_\ell^n\}$ weak $(\star)$ converge to zero.*

COROLLARY 4.26 [**100**] (the duality relation). *For $f \in C(Q)$,*

$$E(f) = \sup_{\ell \subset Q} \overline{\lim_{n\to\infty}}\, r_\ell^n(f), \tag{4.37}$$

where the supremum is taken over all closed or infinite lightning bolts $l \subset Q$.

The main advantage of Corollary 4.26 in comparison with Theorem 4.8 is that here we do not have to "leave" Q for its neighborhood. At the same time, Theorem 4.8 is based on the description of all measures in $D(Q)^\perp$, not merely extreme points.

COROLLARY 4.27. *If $D(Q)$ is closed in $C(Q)$, then*

$$E(f) = \sup_{\ell \subset Q} |r_\ell(f)|, \tag{4.38}$$

where sup *is taken over all closed lightning bolts ℓ in Q.*

PROOF. The assumption concerning closedness of $D(Q)$ according to Medvedev's criterion (Theorem 2.2) implies the possibility of closing up every finite lightning bolt by adding finitely many vertices whose number does not depend on the particular lightning bolt. Considering an infinite lightning bolt ℓ and its finite parts ℓ^n, close up those parts to closed lightning bolts L^n. It is easy to see that $r_{\ell^n} - r_{L^n} \to 0$ in the weak $(\star)$ topology.

PROOF OF THEOREM 1.6 FOR ARBITRARY (NOT NECESSARILY METRIC) COMPACT SPACES. *Necessity.* According to Theorem 4.15, Q cannot contain closed lightning bolts. Now let $\ell = [p_1, \dots, p_n]$ be an irreducible lightning bolt. According to Theorem 2.5 of Chapter 1, there must exist a number λ, $0 < \lambda \le 1$, such that for any measure $\mu \in C(Q)^\star$ we have $\max_{i=1,2} \|\pi_i \circ \mu\| \ge \lambda\|\mu\|$. However, $\|r_\ell\| = 1$, and it is easy to calculate that $\|\pi_i \circ r_\ell\| \le \dfrac{2}{n}$. Hence, $\dfrac{2}{n} \ge \lambda$ and $n \le \dfrac{2}{\lambda}$.

Sufficiency. According to Theorem 2.2 the subspace D is closed. Since there are no closed lightning bolts, all lightning bolts are irreducible and the uniform boundedness of their lengths implies that for any point $p \in Q$ its equivalence class $R(p)$ with respect to the relation R is a closed set. According to Theorem 4.15 we obtain that D is everywhere dense and, therefore, coincides with $C(Q)$.

Examples. To conclude the section let us give a few examples associated with Theorem 4.15 and some of the examples given in Section 2.

EXAMPLES 4.28 (All examples are in $\mathbb{R}^2$).

(a) If Q has interior points, then $\overline{D} \ne C(Q)$ since Q contains closed lightning bolts. Whether $\overline{D} = D$ depends on the geometry of Q. In simple domains, indeed $\overline{D} = D$ as is easily seen from Theorem 2.2. It is equally easy to construct domains containing irreducible lightning bolts of an arbitrarily large length. Hence, by Theorem 2.2, $\overline{D} \ne D$ there.

(b) Generalization: Q has positive area. Here, $\overline{D} \ne C(Q)$. Indeed, we can place inside Q vertices of a square with sides parallel to the coordinate axes. Let us prove this assertion. Almost every point of a set of positive measure is a density point (see [**111**]). Let $p_0 \in Q$ be such a point. Then, there exists a square Δ centered at p_0 with sides parallel to the coordinate axes such that $\sigma(\Delta \cap Q) > \dfrac{3}{4}\sigma(\Delta)$ (σ is the Lebesgue measure on the plane) and hence $\sigma(\Delta \backslash Q) < \dfrac{1}{4}\sigma(\Delta)$. Divide Δ into four equal squares by lines passing through p_0 and parallel to the axes. We obtain the squares $\Delta_1, \dots, \Delta_4$ with centers $p_1, \dots, p_4$. Let Δ_0 be another square congruent to Δ_i, $i = 1, \dots, 4$, and having sides parallel to the coordinate axes. Denote its center by Q. Translate the squares Δ_i, $i = 1, \dots, 4$, so that they coincide with the square Δ_0. Call a point in Δ_0 "bad" if it corresponds to at least one point from

$\bigcup_{i=1}^{4}(\Delta_i \backslash Q)$. The area σ of all "bad" points satisfies

$$\sigma \leq \sum_{1}^{4} \sigma\left(\Delta_i \backslash Q\right)=\sigma(\Delta \backslash Q)<\frac{1}{4} \sigma(\Delta)=\sigma\left(\Delta_0\right).$$

So, there are points in Δ_0 that are not "bad". Let $b \in \Delta_0$ be such a point. Then its preimages b_i in Δ_i, $i=1, \ldots, 4$, belong to Q. Each point b_i is obtained from the center p_i of the square Δ_i by translating it by the same vector $\vec{ab}$. Therefore, $b_1, \ldots, b_4$ are vertices of the required square.

This argument is due to V. A. Medvedev, who also has observed an even more general curious geometric fact. For an arbitrary finite set of points E in the plane we can find in a set Q of positive area a subset $\varepsilon \subset Q$ that is obtained from E by a dilation and translation. Thus, in particular, Q contains closed lightning bolts of any given form.

(c) Let Q be a closed Jordan curve. By Shnirel′man's theorem [**120a**] there exists a square with vertices on Q. If the sides of such a square are parallel to the coordinate axes, then $D \neq C(Q)$. The natural question of whether always $D=\{\varphi(x)+\psi(y)\}$ is not dense in $C(Q)$ has a negative answer (cf. Example 2.10c).

(d) Let Q be a compact in $\mathbb{R}^2$ without interior. Can one find functions $\varphi_1(p) \in C(Q)$ and $\varphi_2(p) \in C(Q)$ so that the subspace

$$\left\{\varphi \circ \varphi_1(p)+\psi \circ \psi_2(p)\right\}, \qquad \varphi \in C\left(\varphi_1(Q)\right), \quad \psi \in C\left(\varphi_2(Q)\right)$$

is everywhere dense in $C(Q)$? If Q is a totally disconnected set, then there exists a simple Jordan arc Γ such that $\Gamma \supset Q$. Let $x=x(t)$, $y=y(t)$ be the equation of that arc. Then on Γ, and therefore on Q, $t=\varphi_1(p)$, $p=(x, y)$, is a continuous function that is injective on Q. Then every $f(p) \in C(Q)$ can be written as $\varphi \circ \varphi_1(p)$, and hence $\left\{\varphi \circ \varphi_1(p)\right\}=C(Q)$. In general, though, the answer to the above question is unknown.

In conclusion, let us prove the following striking result.

THEOREM 4.29. *Let $N=2$. If $\overline{BD}=B(Q)$, then $BD=B(Q)$.*

PROOF. There are no closed lightning bolts in Q, since for any such bolt ℓ we have $r_\ell \in D^{\perp}$. Therefore, all finite lightning bolts are irreducible and any two points in Q can be joined by at most one lightning bolt. For every equivalence class $E \in Q/R$, mark and fix a starting point p_1^E. A point p, as in the proof of Theorem 8.1 in Section 1 of Chapter 1, is called a point of rank n if the lightning bolt joining p with the starting point p_1^E has length n. Denote by Q_1 the set of all points in Q whose rank is odd, and by Q_2 the set of points of even rank. Then $Q_1 \cap Q_2=\emptyset$ and $Q_1 \cup Q_2=Q$. In view of Theorem 1.5 we need to show that the lengths of all lightning bolts are bounded by a number that does not depend on the particular lightning bolt. If it is false, then for any natural integer n there exists a lightning bolt $[p_1, p_2, \ldots, p_n]$, where $p_1=p_1^E$ for some $E \in Q/R$. Define a function

$$F(p)=1, \quad p \in Q_1, \qquad F(p)=-1, \quad p \in Q_2.$$

Since $\overline{BD}=B(Q)$, there exists a function $f(p)$ satisfying

$$f(p)=g_1 \circ \pi_1(p)+g_2 \circ \pi(p), \qquad g_1 \in B(X), \quad g_2 \in B(Y),$$

and such that

$$f(p) > \frac{1}{2} \text{ for } p \in Q_1, \qquad f(p) < -\frac{1}{2} \text{ for } p \in Q_2. \tag{4.39}$$

Denote $h_1 = g_1 \circ \pi_1$, $h_2 = g_2 \circ \pi_2$. Then

$$f = h_1 + h_2. \tag{4.40}$$

Assume, for example, that in our lightning bolt $p_1 R_1 p_2$. Then,

$$h_1(p_2) = h_1(p_1), \qquad h_2(p_3) = h_2(p_2), \qquad h_1(p_4) = h_1(p_3),$$

etc. From (4.39) we obtain

$$\begin{aligned}
&h_1(p_1) = f(p_1) - h_2(p_1);\\
&h_2(p_2) = f(p_2) - h_1(p_2) = f(p_2) - f(p_1) + h_2(p_1);\\
&h_1(p_3) = f(p_3) - h_2(p_3) = f(p_3) - f(p_2) + f(p_1) - h_2(p_1);\\
&h_2(p_4) = f(p_4) - h_1(p_4) = f(p_4) - f(p_3) + f(p_2) - f(p_1) + h_2(p_1);
\end{aligned}$$

and so on. Therefore, for $2k \le n$, we have:

$$\begin{gathered}
h_2(p_{2k}) = f(p_{2k}) - f(p_{2k-1}) + f(p_{2k-2}) - \cdots - f(p_1) + h_2(p_1),\\
h_2(p_1) - h_2(p_{2k}) = f(p_1) - f(p_2) + f(p_3) - \cdots - f(p_{2k}).
\end{gathered}$$

From this and (4.40) it follows that

$$h_2(p_1) - h_2(p_{2k}) > k,$$

which contradicts the boundedness of h_2. Hence, the lengths of lightning bolts in Q are uniformly bounded. Also, Q contains no closed lightning bolts. Theorem 1.5 shows that $BD = B(Q)$.

Theorem 4.29 is due to V. A. Medvedev (unpublished).

§5. Relation to the theory of functional equations

Let's show some applications of the above theory to the study of functional equations. This possibility was discovered by Buck [**26–28**] and stimulated the study of the above problems. The main results presented in §§1–4 had not yet been discovered at that time.

Let $k(x)$, $\beta(x)$ and $u(x)$ belong to $C[0,1]$, and let $\beta(x) : [0,1] \to [0,1]$. Consider a functional equation

$$\varphi(x) - k(x)\varphi(\beta(x)) = u(x) \tag{5.1}$$

with an unknown function $\varphi \in C[0,1]$. The equation (5.1) has been studied by many authors. A survey of results can be found in [**90a**]. The case $\|k\| < 1$ is simple, since one can use the principle of contracting mappings. Our theory relates to the case $k = 1$. Following [**26**], consider a more general situation.

Let X and Y be compact spaces and $\beta_0(x), \dots, \beta_n(x)$ be continuous mappings of X into Y. Given functions $u_k(x) \in C(X)$, $k = 1, \dots, n$, we look for a function $\varphi \in C(Y)$ satisfying a system of functional equations:

$$\begin{aligned} &\varphi(\beta_0(x)) - \varphi(\beta_1(x)) = u_1(x); \\ &\dots\dots\dots\dots\dots \\ &\varphi(\beta_{n-1}(x)) - \varphi(\beta_n(x)) = u_n(x). \end{aligned} \tag{5.2}$$

A necessary condition for solvability. Set

$$\Gamma_{i,j} = \{x : \varphi_i(x) = \varphi_j(x)\}, \qquad 0 \le i < j \le n. \tag{5.3}$$

LEMMA 5.1. *In order that the system* (5.2) *have a solution it is necessary and sufficient that the right-hand sides of* (5.2) *satisfy the following relations*:

$$\begin{aligned} u_k = 0 \quad &\text{for } x \in \Gamma_{k-1,n}, \qquad k = 1, \dots, n \\ u_k(x) + \cdots + u_m(x) = 0 \quad &\text{for } x \in \Gamma_{n-1,m}, \qquad 1 \le k \le m \le n. \end{aligned} \tag{5.4}$$

PROOF. From the k-th equation in (5.2) we obtain that $u_k(x) = 0$ at those points where $\beta_{k-1}(x) = \beta_k(x)$. Adding equations with indices from k to m, we obtain

$$\varphi(\beta_{k-1}(x)) - \varphi(\beta_m(x)) = u_k(x) + \cdots + u_m(x),$$

which implies all of the relations (5.3).

Approximate solutions. We shall say that the system (5.2) has an approximate solution if for every $\varepsilon > 0$ there exists $\varphi \in C(Y)$ such that

$$\|\varphi(\beta_{k-1}(x)) - \varphi(\beta_k(x)) - u_k(x)\| \le \varepsilon, \qquad k = 1, \dots, n. \tag{5.5}$$

If we can take $\varepsilon = 0$, then φ is an ordinary, exact solution.

Buck's Theorem. In the space $X \times Y$ consider graphs of the mappings $\beta_k(x)$ by setting

$$Q_k = \{(x, \beta_k(x)), x \in X\}, \qquad k = 0, 1, \dots, n; \quad Q = Q_0 \cup \cdots \cup Q_n. \tag{5.6}$$

THEOREM 5.2. *In order that the system* (5.2) *have a solution for arbitrary right-hand sides* $u_1(x), \dots, u_n(x)$ *that satisfy* (5.4), *it is necessary and sufficient that* $D(Q) = C(Q)$. *In order that the system* (5.2) *with arbitrary right-hand sides* $u_k(x)$ *satisfying* (5.4) *have an approximate solution, it is necessary and sufficient that* $D(Q)$ *be everywhere dense in* $C(Q)$ (*i.e.*, $\overline{D(Q)} = C(Q)$).

PROOF. Let $u_1(x), \dots, u_n(x)$ belong to $C(X)$ and satisfy (5.4). Define a function F on Q by setting

$$F(x,y) = \begin{cases} 0, & x \in Q_0 \\ -\sum_1^k u_j(x), & x \in Q_k, \qquad 1 \le k \le n. \end{cases} \tag{5.7}$$

(5.4) guarantees that $F(x, y)$ is well-defined on Q and is continuous.

Suppose that $D(Q)$ is everywhere dense in $C(Q)$. Then, for every $\varepsilon > 0$ there exists $f(x,y) = A(x) + B(y)$ so that

$$\|F - f\|_Q < \varepsilon. \tag{5.8}$$

Hence,

$$\begin{aligned} &\|A(x) + B(\beta_0(x)) - 0\| \le \varepsilon, \\ &\|A(x) + B(\beta_1(x)) + u_1(x)\| \le \varepsilon, \\ &\left\| A(x) + B(\beta_k(x)) + \sum_1^k u_j(x) \right\| \le \varepsilon, \qquad 1 \le k \le n. \end{aligned} \tag{5.9}$$

From the first two inequalities in (5.9) we obtain

$$\|B(\beta_0(x)) - B(\beta_1(x)) - u_1(x)\| \le 2\varepsilon.$$

Similarly, from the k-th and $(k-1)$-st inequalities it follows that

$$\|B(\beta_{k-1}(x)) - B(\beta_k(x)) - u_k(x)\| \le 2\varepsilon.$$

Therefore, the function $B(y) \in C(Y)$ is an approximate solution of the system (5.2). If $D(Q) = C(Q)$, then taking $\varepsilon = 0$ we obtain as before that $B(y)$ is an exact solution of (5.2).

Let us now assume that the system (5.2) has an approximate solution. We need to show that $\overline{D(Q)} = C(Q)$. Suppose there is a function $F(x, y) \in C(Q)$ defined on Q. Set

$$u_k(x) = F(x, \beta_{k-1}(x)) - F(x, \beta_k(x)). \tag{5.10}$$

A straightforward calculation yields that all such $u_k(x)$ satisfy conditions (5.4). Let $\varphi(y)$ be an approximate solution of the system (5.2). Consider the function

$$f(x, y) = f(x, \beta_0(x)) - \varphi(\beta_0(x)) + \varphi(y) \in D(Q). \tag{5.11}$$

On Q_0, we have

$$f(x, y) = f(x, \beta_0(x)) = F(x, y).$$

On Q_1, using (5.10), we have

$$\begin{aligned} f(x, y) =& f(x, \beta_1(x)) = F(x, \beta_0(x)) - \varphi(\beta_0(x)) + \varphi(\beta_1(x)) \\ =& F(x, \beta_0(x)) - u_1(x) - [\varphi(\beta_0(x)) - \varphi(\beta_1(x)) - u_1(x)] \\ =& F(x, \beta_1(x)) - [\varphi(\beta_0(x)) - \varphi(\beta_1(x)) - u_1(x)]. \end{aligned}$$

Hence,

$$\|f - F\|_{Q_1} < \varepsilon. \tag{5.12}$$

In general, for $1 \le k \le n$ we have on Q_k

$$\begin{aligned} f(x, y) =& F(x, \beta_0(x)) - \varphi(\beta_0(x)) + \varphi(\beta_k(x)) = [F(x, \beta_0(x)) - F(x, \beta_1(x))] \\ &+ [F(x, \beta_1(x)) - F(x, \beta_2(x))] + \cdots + [F(x, \beta_{k-1}(x)) - F(x, \beta_k(x))] \\ &+ F(x, \beta_k(x)) - [\varphi(\beta_0(x)) - \varphi(\beta_1(x))] - [\varphi(\beta_1(x)) - \varphi(\beta_2(x))] \\ &- \cdots - [\varphi(\beta_{k-1}(x)) - \varphi(\beta_k(x))] = u_1(x) + \cdots + u_k(x) + F(x, \beta_k(x)) \\ &- [\varphi(\beta_0(x)) - \varphi(\beta_1(x))] - \cdots - [\varphi(\beta_{k-1}(x)) - \varphi(\beta_k(x))]. \end{aligned}$$

From the latter we obtain

$$\begin{aligned} \|f - F\|_{Q_k} &= \|f(x, \beta_k(x)) - F(x, \beta_k(x))\| \\ &\le \sum_1^k \|\varphi(\beta_{j-1}(x)) - \varphi(\beta_j(x)) - u_j(x)\| < k\varepsilon. \end{aligned} \tag{5.13}$$

From (5.11)–(5.13) we obtain that $\overline{D(Q)} = C(Q)$. If the system (5.2) had an exact solution $\varphi(y)$, then performing the same calculations with $\varepsilon = 0$, we would show that $D(Q) = C(Q)$. The proof is complete.

A special case.

COROLLARY 5.3 [**26**]. *Let $p(x)$ be an increasing continuous function on $[0,1]$, $\beta(0) = 0$, $\beta(1) = 1$, and $0 < \beta(x) < x$ for $0 < x < 1$. The functional equation*

$$\varphi(x) - \varphi(\beta(x)) = u(x) \tag{5.14}$$

with an arbitrary $u(x) \in C[0,1]$ has an approximate solution satisfying $u(0) = u(1) = 0$.

PROOF. In our case, $\beta_0(x) = x$, $\beta_1(x) = \beta(x)$, and $\Gamma_{0,1} = \{0,1\}$. Therefore, the necessary requirements (5.4) are reduced to $u(0) = u(1) = 0$. The set Q is a union of graphs $Q_0 : y = x$ and $Q_1 : y = \beta(x)$. Since Q contains lightning bolts with an arbitrarily large number of vertices, Theorem 2.2 yields that $D(Q)$ is not closed and so $D(Q) \neq C(Q)$. Yet, $D(Q)^{\perp} = \{0\}$ (cf. Example 2.10(c)) and, accordingly, $\overline{D(Q)} = C(Q)$. Therefore, equation (5.14) need not have an exact solution for all $u(x)$ satisfying the necessary requirements $u(1) = u(0) = 0$; nevertheless, for any such $u(x)$ there exists an approximate solution.

Note that in [**26**], due to the lack of criteria needed to make direct statements concerning the place of $D(Q)$ inside $C(Q)$, the conclusions were presented in the reverse order: first, by some special means, it was shown that (5.4) has an approximate solution, and from that the conclusions about $D(Q)$ were derived:

An application to the moment problem.

COROLLARY 5.4 [**26**]. *Let $\beta(x)$ be the same as in Corollary 5.3. If for a Borel, real-valued measure μ on $[0,1]$ we have*

$$\int_0^1 [x^n - (\beta(x))^n]\, d\mu = 0, \qquad n = 1, 2, \dots, \tag{5.15}$$

then $S_\mu = \{0,1\}$. In particular, if $\mu_n = \int_0^1 x^n d\mu$, then $\mu_1 = \mu_2 = \cdots$.

PROOF. By the Weierstrass approximation theorem, (5.15) implies that for any $\varphi(x) \in C[0,1]$ we must have $\int_0^1 [\varphi(x) - \varphi(\beta(x))]\, d\mu = 0$. But the equation (5.14) has an approximate solution for all $u(x)$ such that $u(1) = u(0) = 0$. For any $v(x) \in C([0,1])$, the function $u(x) = x(x-1)v(x)$ satisfies $u(0) = u(1) = 0$. Hence $d\mu_1 = x(x-1)d\mu = 0$, and therefore μ can only have atoms at $x = 0$ and $x = 1$.

§6. Chebyshev-like problems for the best approximation of a function of two variables by sums $\varphi(x) + \psi(y)$

Chebyshev's ideas in the theory of best approximation. By Chebyshev's ideas in the theory of best approximation, we generally understand (cf., e.g., [**2**]) the study of characteristic properties of functions that give the best approximation, problems of their uniqueness, relations to other extremal problems (duality), methods of calculating or estimating the quantity of the best approximation, and algorithms allowing one to construct functions giving the best approximation. Questions concerning existence of the best approximation should not, strictly speaking, be included into the circle of "Chebyshev's ideas", since P.L. Chebyshev did not himself consider such questions (in relation to polynomials or rational functions), taking existence of functions giving the best approximation for granted. However, in the case of infinite-dimensional subspaces the question is non-trivial. In this section we add a few existence results to those of §3. The following section deals with algorithms.

Chebyshev's Theorem [79]. Let X, Y, and Q be compact spaces, $Q \subset X \times Y$, and let π_1 and π_2 be natural projections from $X \times Y$ onto X and Y, respectively. Then $D = D(Q)$ is the subspace in $C(Q)$ that consists of functions $\varphi(x) + \psi(y)$, $\varphi(x) \in C(X)$, $\psi(y) \in C(Y)$. For an arbitrary $f \in C(Q)$, set

$$E(f) = \operatorname{dist}(f, D) = \inf_{\substack{\varphi \in C(X) \\ \psi \in C(Y)}} \|f(p) - \varphi(x) - \psi(y)\|_Q, \qquad p = (x, y) \in Q. \tag{6.1}$$

So, we are considering the best approximation of $f(p)$ by functions from the subspace D. Functions $\varphi^\star(x) + \psi^\star(y) \in D(Q)$ realizing the inf in (6.1) are called best approximations of f in $D(Q)$. (We assume that at least one such function exists.) In §2, while studying properties of the subspace D, we considered in passing simpler problems of approximation of $f(x, y)$ by functions of only one variable—problems (2.2)–(2.12), (2.17), (2.21).

THEOREM 6.1. *In order that the sum $\varphi^\star(x) + \psi^\star(y) \in D(Q)$ give the best approximation of $f(x, y) \in C(Q)$ among all functions in $D(Q)$, it is necessary and sufficient that there exist a lightning bolt $\ell \subset Q$ with the following properties:*

1. *ℓ is either closed or infinite.*
2. *At the vertices of ℓ the expression $f(x, y) - \varphi^\star(x) - \psi^\star(y)$ assumes values $\pm M$, where $M = \|f - \varphi^\star - \psi^\star\|$ and, moreover, the signs at the neighboring points are opposite.*

Thus, in our problem of the best approximation, which is quite distant from approximation by polynomials, the celebrated Chebyshev alternation appears in a peculiar but clear form.

Before giving a proof of the theorem, let us recall the general criterion for the best approximation (cf., e.g., [**120**]) of real-valued, continuous functions on a compact Q in the space $C(Q)$.

THEOREM 6.2. *Let E be a subspace in $C(Q)$. In order that $\Phi^\star(p) \in E$ be the function of the best approximation to a given function $f(p) \in C(Q)$, it is necessary and sufficient that there exist a Borel real-valued measure $\mu^\star \in E^\perp$ such that*

$$\begin{gathered} \|\mu^\star\| = \int_Q |d\mu^\star| = 1, \qquad \mu^\star = \mu^{\star+} - \mu^{\star-}, \\ f(P) = \Phi^\star(P) = M, \qquad p \in S\left(\mu^{\star+}\right), \\ f(p) - \Phi^\star(p) = -M, \qquad p \in S\left(\mu^{\star-}\right), \quad M = \|f - \Phi^\star\|. \end{gathered} \tag{6.2}$$

Here, as usual, $\mu^{\star+}$ and $\mu^{\star-}$ are components in the Jordan decomposition of $\mu^\star$, $S\left(\mu^{\star+}\right)$ and $S\left(\mu^{\star-}\right)$ are their closed supports, $\|\mu^\star\|$ is the total variation of $\mu^\star$. For $\mu^\star$ one can take any measure for which the supremum in the duality relation (4.11) is attained and which is normalized by the condition $\int_Q f d\mu^\star > 0$. In addition, any charge $\mu^\star$, realizing the above-mentioned supremum, characterizes via (6.2) any function $\Phi^\star$ of the best approximation.

We shall need a lemma that is closely related to Lemma 4.18, but is much easier.

LEMMA 6.3. *Let a measure $\mu \in D(Q)^\perp$, and let $\mu = \mu^+ - \mu^-$ be its Jordan decomposition. If $S(\mu^+)$ and $S(\mu^-)$ do not overlap, then $S(\mu)$ contains either a closed lightning bolt ℓ or an infinite one. In addition, the vertices of ℓ with odd numbers are in $S(\mu^+)$, while those with even numbers are in $S(\mu^-)$.*

PROOF. Let $p_1 \in S(\mu^\star)$. Consider all neighborhoods G of p_1. For every neighborhood G, according to Lemma 4.1, we have

$$\mu\left(\pi_1^{-1} \circ \pi_1(G)\right) = 0.$$

Therefore, in $\pi_1^{-1} \circ \pi_1(G)$ there must exist a point $q_G \in S(\mu^-)$ and a point $p_G \in G$, such that $\pi_1(p_G) = \pi_1(q_G)$. The set of all neighborhoods $\{G\}$ is a directed set, while the net of points $\{p_G\} \to p_1$. From the net $\{q_G\}$ we can choose a subnet $\{q_\beta\}$ that converges to a point $p_2 \in S(\mu^-)$, so $p_2 \neq p_1$. We have $\pi_1(p_2) = \lim \pi_1(q_\beta) = \lim \pi_1(p_\beta) = \lim \pi_1(p_G) = \pi_1(p_1)$. Similarly, starting with p_2 and using the projection π_2 we construct a point $p_3 \in S(\mu^+)$ for which $\pi_2(p_2) = \pi_2(p_3)$, etc. Either the resulting lightning bolt $[p_1, p_2, p_3 \ldots]$ closes down at a certain step, or it is infinite.

PROOF OF THEOREM 6.1. *Necessity.* Let $\varphi^\star(x) + \psi^\star(y)$ be the best approximation to f and $\mu^\star$ be a charge characterizing it as in Theorem 6.2. According to Theorem 6.2, $S(\mu^{\star+})$ is contained in a set where $f - \varphi^\star - \psi^\star = M$ while $S(\mu^{\star-})$ is contained in one where $f - \varphi^\star - \psi^\star = -M$. Hence, $S(\mu^{\star+})$ and $S(\mu^{\star-})$ do not overlap and in $S(\mu^\star)$ there exists a lightning bolt ℓ satisfying the required properties.

Sufficiency. First, let $\ell = [p_1, \ldots, p_n]$ be a closed lightning bolt satisfying all the properties listed in the theorem (n, then, is an even number). Define a functional r^ℓ by the formula (4.6) ($r_\ell \in D^\perp$ according to Lemma 4.5). Clearly, for the measure $\mu^\star = r_\ell$ defining the functional all relations (6.2) are satisfied. (For the sake of definiteness, we assume that $(f^\star - \varphi^\star - \psi^\star)(p_1) = M$.) Now let the lightning bolt $\ell = [p_1, \ldots, p_n, \ldots]$ be infinite. Define a sequence of functionals $\{r_\ell^n\}$, $n = 1, 2, \ldots$, as in Lemma 4.5. Then

$$\int_a dr_\ell^{n+} \to \frac{1}{2}, \qquad \int_Q dr_\ell^{n-} \to \frac{1}{2}.$$

We can assume that the sequence $\{r_\ell^{n+}\}$ converges weak ($\star$) to a measure μ^+ while $\{r_\ell^{n-}\}$ converges weak ($\star$) to a measure μ^- and, in addition,

$$\int_Q d\mu^+ = \frac{1}{2}, \qquad \int_Q d\mu^- = \frac{1}{2}.$$

According to Lemma 4.5, the functional $\mu^\star$ defined by the measure $\mu^\star = \mu^+ - \mu^-$ belongs to the annihilator $D^\perp$. The closed support $S(\mu^+)$ of μ^+ belongs to the set of limit points for the set of vertices of the lightning bolt with odd numbers (again, for the sake of definiteness, we suppose that $(f - \varphi^\star - \psi^\star)(p_1) = M$, while $S(\mu^-)$ is situated in the set of limit points for the set of vertices with even indices). Therefore, it is clear that at the points of $S(\mu^\star)$ the values of $f - \varphi^\star - \psi^\star$ equal M while at points of $S(\mu^-)$ the values of $f - \varphi^\star - \psi^\star$ equal $-M$ (in particular, $S(\mu^+)$ and $S(\mu^-)$ do not overlap and hence $\mu^+ - \mu^-$ is the Jordan decomposition for $\mu^\star$). The proof is now complete since all the requirements of (6.2) are fulfilled ($\|\mu^\star\| = \int_Q d\mu^+ + \int_Q d\mu^- = 1$).

Levelling of a function. A function $f(x, y)$ given on $Q \subset X \times Y$ is said to be levelled with respect to y if

(6.3) $$\sup_y f(x, y) = -\inf_y f(x, y) \quad \text{for all } x \in X,$$

and levelled with respect to x if

(6.4) $$\sup_x f(x, y) = -\inf_x f(x, y) \quad \text{for all } y \in Y.$$

If $f(x, y)$ is levelled with respect to both variables, we shall simply call it levelled.

COROLLARY 6.4. *If $f(x, y) \in C(Q)$ is levelled, then $E(f) = \|f\|$.*

Indeed, let p_1 be a point where $f(p_1) = \|f\|$. Since f is levelled there exists a point p_2 such that $\pi_1(p_1) = \pi_1(p_2)$, $f(p_2) = -\|f\|$. Again, since f is levelled, there exists a point p_3 such that $f(p_3) = \|f\|$, $\pi_2(p_2) = \pi_2(p_3)$, etc. We obtain a lightning bolt $[p_1, p_2, p_3, \dots]$ with all the properties as required by Theorem 6.1. So, if for some $f \in C(Q)$ we choose a function $\varphi^\star(x) + \psi^\star(y) \subset D(Q)$ so that $f - \varphi^\star - \psi^\star$ is a levelled function, then $\varphi^\star(x) + \psi^\star(y)$ is the best approximation of f in $D(Q)$.

There exists a simple algorithm that allows us, for a given function f, to construct a levelled function $f(x, y) - \varphi^\star(x) - \psi^\star(y)$. It is given in §7. If $f(x, y) - \varphi^\star(x) - \psi^\star(y)$ is a levelled function, then $\varphi^\star(x) + \psi^\star(y)$ is not merely the best approximation to f but, in a certain sense, "the best of the best" approximations. This, too, is discussed below.

Realization of either one of two situations in the Chebyshev criterion. Both possibilities mentioned in Theorem 6.1, the case of a closed lightning bolt and that of a bolt containing infinitely many links, may indeed occur. As to closed lightning bolts, this is obvious. To construct an example for which the second possibility occurs is more difficult, since the lengths of links of a polygonal path cannot tend to zero; indeed, at the end-points of each link the expression $f - \varphi^\star - \psi^\star$ assumes values of opposite sign but equal modulus.

The first example of such a phenomenon was constructed in the author's paper [**79**]. Below we shall present that and other similar constructions. However, first let us establish a simple sufficient criterion for existence of a closed lightning bolt satisfying the properties of Theorem 6.1.

PROPOSITION 6.5. *Let $\ell = [p_1, p_2 \dots]$ be an infinite lightning bolt from Theorem 6.1, yet the set of its vertices has only finitely many limit points. In that case, there exists a closed lightning bolt $L = [q_1, \dots, q_{2m}]$ such that $f(q_i) - \varphi^\star(q_i) - \psi^\star(q_i) = (-1)^{i-1} E(f)$.*

PROOF. Consider the measure $\mu^\star$ obtained in the proof of Theorem 6.1 for the case of an infinite lightning bolt. Clearly, the closed support of $\mu^\star$ belongs to the set of the limit points for the collection of vertices of the lightning bolt ℓ. Starting with a point $q_1 \in S(\mu^{\star +})$ and repeating the construction in the proof of necessity in Lemma 6.3 and Theorem 6.1, we necessarily arrive at a closed lightning bolt because $S(\mu^\star)$ is a finite set.

EXAMPLES. (All constructions are in $\mathbb{R}^2$.)

EXAMPLE 6.6. We need to construct two disjoint closed sets Φ_1 and Φ_2 with the following properties:

(a) Φ_1 and Φ_2 have equal projections on the coordinate axes;
(b) any lightning bolt whose vertices lie in Φ_1 and Φ_2 so that two neighboring vertices (along the bolt) cannot belong to the same set, has infinitely many links (cannot be closed).

The fact that projections of Φ_1 and Φ_2 on the coordinate axes are equal provides existence of a lightning bolt with vertices in Φ_1 and Φ_2 and the alteration described above (we can start the lightning bolt at any point in Φ_1 or Φ_2).

CONSTRUCTION OF AN EXAMPLE. We shall approach the desired set $\Phi = \Phi_1 \cup \Phi_2$ in several steps. At the first step consider the main rectangle U_1 with vertices A_1^1, A_2^1, A_3^1, A_4^1 (with sides parallel to the coordinate axes and, in addition, the side $A_1^1A_2^1$ parallel to Ox). At the second step we construct the lightning bolt U_2 with eight vertices A_i^2, $i = 1, \dots, 8$, as follows. Surround points A_1^1, A_2^1, A_3^1, A_4^1 by disjoint small neighborhoods $S_1^1, \dots, S_4^1$ that we assume to be closed. Taking in S_1^1 a point A_1^2, issue from it a line parallel to Ox until we arrive at S_2^1. Taking for A_2^2 one of the points of intersection of our line with S_2^1, issue from A_2^2 a line parallel to Oy until the intersection with S_3^1. In S_3^1 choose vertex A_3^2 as before. From A_3^2 issue a link parallel to Ox until the intersection with S_4^1, where we choose vertex A_4^2, but so that the line $A_4^2A_4^3$ is not parallel to Oy. From A_4^2 issue a link parallel to Oy until the intersection with S_4^1, where we choose a vertex A_5^2 and then repeat circuiting the main rectangle U_1 constructing vertices A_6^2, A_7^2, and A_8^2, the latter so that the vertical link $A_8^2A_1^2$ closes the lightning bolt. During the construction we also watch for all the links of U_1 and U_2 to run along different lines. Surround the points A_i^2, $i = 1, \dots, 8$, by disjoint closed neighborhoods S_i^2, $i = 1, \dots, 8$, every one of them lying in a corresponding neighborhood S_i^1, and construct a lightning bolt U_3 with sixteen vertices by a process similar to that used for U_2.

Suppose at the k-th step we have constructed a closed lightning bolt U_k with the vertices $A_1^k, \dots, A_{\nu_k}^k$ (the number of vertices $\nu_k = 4 \cdot 2^{k-1}$). Surround the point A_i^k, $i = 1, \dots, \nu_k$, by disjoint neighborhoods S_i^k, every one of which lies in a corresponding neighborhood S_i^{k-1} that appeared at a previous step of the construction. Then, to construct U_{k+1} with the number of vertices $4 \cdot 2^k$ we proceed as follows. Choose a vertex A_1^{k+1} in S_1^k and run from it a link parallel to the link $A_1^kA_2^k$ (i.e., parallel to the Ox-axis) till the intersection with S_2^k. Choose a vertex A_2^{k+1} in S_2^k. Then, from A_2^{k+1} run a link parallel to $A_2^kA_3^k$ and choose a vertex A_3^{k+1} in S_3^k, etc. Choose a vertex $A_{4\cdot 2^{k-1}}^{k+1}$ in $S_{4\cdot 2_{k-1}}^k$ such that the segment $A_{4\cdot 2^{k-1}}^{k+1}A_1^{k+1}$ is not parallel to the link $A_{4\cdot 2^{k-1}}^kA_1^k$ (i.e., the Oy-axis). From $A_{4\cdot 2^{k-1}}^{k+1}$ run a link parallel to Oy and the next vertex $A_{4\cdot 2^{k-1}}^{k+1}$ chosen in S_1^k. Starting with $A_{4\cdot 2^{k-1}+1}^{k+1}$, repeat circuiting the lightning bolt U_k along links parallel to the corresponding links of U_k. Choose a vertex $A^{k+1}_{4\cdot 2^k}$ in $S^k_{4\cdot 2^{k-1}}$ so that the link $A^{k+1}_{4\cdot 2^k}A^{k+1}_1$ would close U_{k+1}. While constructing all the links of U_{k+1} we also secure that all the links of all the lightning bolts $U_1, \dots, U_{k+1}$ belong to different lines. This allows us to choose at the k-th step neighborhoods $S^k_1, \dots, S^k_{4\cdot 2^{k-1}}$ so small that any closed lightning bolt whose vertices we only allowed to belong to the set $S^k_1 \cup \dots \cup S^k_{4\cdot 2^{k-1}}$ would have at least ν_k $(\nu_k = 4 \cdot 2^{k-1})$ vertices. It is clear that such a choice of neighborhoods $S^k_1, \dots, S^k_{\nu_k}$ is always possible.

Continue the process of constructing the lightning bolts U_k indefinitely. Thus, we have constructed a sequence of closed lightning bolts $U_1, \dots, U_k, \dots$. Set $S^k = \bigcup\limits_{i=1}^{k} S^k_i$, $\Phi = \bigcap\limits_{k=1}^{\infty} S^k$. Obviously, Φ is a closed set since all the neighborhoods S^k_i are closed. Denote by Φ_1 the part of Φ that belongs to $S^1_1 \cup S^1_3$, and by Φ_2 the part that belongs to $S^1_1 \cup S^1_4$. Let us show that Φ_1 and Φ_2 have the same projections on the coordinate axes. (As was mentioned above, this would show the existence of lightning bolts whose vertices alternate between Φ_1 and Φ_2.) Let $A \in \Phi_1$. For definiteness, assume that $A \in S^1_2$. There exists a sequence of closed spheres $S^1_1 \supset S^2_{i_2} \supset S^3_{i_3} \supset \dots$ of decreasing ranks shrinking to the point A. By our construction the sequence $S^1_2 \supset S^2_{i_2+1} \supset S^3_{i_3+1} \supset \dots$ shrinks to a point $B \in \Phi_2$ having the same ordinate as A. Similarly, we can show the existence of a point $C \in \Phi_2$ that has the same abscissa as A. It remains to show that there are no closed lightning bolts whose vertices alternate between Φ_1 and Φ_2.

Again, let A be a point in Φ; without loss of generality, assume again that it belongs to S^1_1. Let us show that every lightning bolt starting at A and having its vertices alternate between Φ_1 and Φ_2 cannot close and, therefore, must contain infinitely many links. Consider the neighborhoods $S^k_1, \dots, S^k_{4\cdot 2^{k-1}}$ constructed at the k-th step of our procedure. The set Φ belongs to the sum of those neighborhoods and has a nonempty intersection with every one of them. Assume that a lightning bolt starting at A is closed. Then its vertices, while belonging to Φ must also belong to $S^k = S^k_1 \cup \dots \cup S^k_{4\cdot 2^{k-1}}$ and, in view of the property of neighborhoods S^k_i mentioned above, the number of those vertices cannot be less than $4 \cdot 2^{k-1}$. Since this holds for all k and $4 \cdot 2^{k-1} \to \infty$, our lightning bolt cannot be closed.

Now, taking a large rectangle $F \supset \Phi$, construct a function $f \in C(F)$ such that

$$f(x,y) = \begin{cases} 1, & (x,y) \in \Phi_1, \\ -1, & (x,y) \in \Phi_2, \\ |f(x,y)| < 1, & (x,y) \in F \backslash \Phi. \end{cases}$$

From Theorem 6.1 and the properties of the set Φ it follows that $E(1) = 1$ and the best approximation is given by the function $\varphi^\star(x) + \psi^\star(y) = 0$, so the lightning bolt mentioned in Theorem 6.1 is infinite.

Let us give another construction of such an example [**100**].

EXAMPLE 6.7. Construct Φ as a set-theoretic limit of compact sets Φ_n. The set Φ_1 consists of four segments L_1, L_2, L_3, L_4 that form a 45° angle with the x-axis and are such that $\pi_1(L_1) = \pi_1(L_2)$, $\pi_1(L_3) = \pi_1(L_4)$, $\pi_2(L_1) = \pi_2(L_4)$, $\pi_2(L_2) = \pi_2(L_3)$. To obtain Φ_n from Φ_{n-1}, turn one of the segments in Φ_{n-1} by 90°. Having done that, remove from all the segments (including the one turned) the middle third. Clearly, Φ_n consists of $4 \cdot 2^{n-1}$ linear segments, and it is easy to check that the equivalence class (with respect to R) of every point in Φ_n consists of $4\cdot 2^{n-1}$ points, one on each segment. The rest follows as in the previous example.

On Φ we can define a measure $\mu \neq 0$, $\mu \in D(\phi)^{\perp}$. For that we construct measures μ_n on linear segments that are contained in Φ_n by starting with the linear measure on those segments normalized so that $\|\mu_n\| = 1$, choosing a positive sign for μ_n on those segments that came from L_1 and L_3 by our process, and a negative sign for those that arose from L_2 and L_4. Unlike in the situation in Proposition 6.4, here $S(\mu)$ consists of infinitely many points and does not contain a closed lightning bolt.

I want to note that in 1969 when I presented Example 6.6 at Professor B. S. Mityagin's seminar, he suggested a modification that essentially coincides with Example 6.7.

The following result further "regularizes" construction of examples similar to 6.6 and 6.7. It is plausible that it has been inspired by Buck's arguments presented in §5.

PROPOSITION 6.8 [**61**]. *Let $T \subset X$ be a compact set for which there exists a homeomorphism $h : T \mapsto T$ such that no power of it (in the sense of superpositions) has fixed points. Let the function $f(x,y) \in C(X\times Y)$ be such that $\|f\| = 1$ and*

$$\begin{aligned} f(x,x) &= 1, \qquad x\in T;\\ f\left(x,h(x)\right) &= -1, \qquad x\in T;\\ |f(x,y)| &< 1, \qquad \textit{elsewhere}. \end{aligned} \tag{6.5}$$

Then, there is no lightning bolt ℓ for which $|r_\ell(f)| = 1$.

PROOF. Let $Q_1 = \{(x,x), x\in T\}$, $Q_2 = \{(x,h(x)), x\in T\}$. Clearly, Q_1 and Q_2 are compact and $Q_1\cap Q_2 = \emptyset$ since h does not have fixed points. Construct a lightning bolt $\ell = [p_1,p_2,p_3,\dots]$ such that $p_{2i-1}\in Q_1$, $p_{2i}\in Q_2$, $\pi_1(p_{2i-1}) = \pi_1(p_{2i})$, $\pi_2(p_{2i}) = \pi_2(p_{2i-1})$. ℓ is infinite, since no power of h has fixed points. We have $f(p_i) = (-1)^{i-1}$; hence, according to Theorem 6.1, $E(f) = 1$, i.e., $\varphi^\star(x) + \psi^\star(y) = 0$ gives the best approximation to f. Any lightning bolt L for which $|r_L(f)| = 1$ must start either at a point in Q_1 or at a point in Q_2 and have its vertices in Q_1 and Q_2, whereas no two neighboring vertices can belong to the same of these two sets. Hence L must be infinite.

EXAMPLE 6.9 [**61**]. Let T be the Cantor ternary set in $X = [0,1]$. Let $x\in T$, $x\neq 1$. Then,

$$x = \sum_{i=1}^{\infty} t_i 3^{-i}, \qquad t_i\in\{0,2\}.$$

Let t_n be the first 0 among $t_1,t_2,\dots$. Set

$$h(x) = \sum_{i=1}^{n-1}(2-t_i)\,3^{-i} + \sum_{n}^{\infty} t_i 3^{-i}, \qquad h(1) = 0.$$

Then all powers of the homeomorphism h have no fixed points.

Proposition 6.8 and Example 6.9 also show that the assumption concerning closedness of R-equivalence classes in Theorem 4.17 was essential. Indeed, consider the restriction of a function f in (6.5) to the set $Q = Q_1\cup Q_2$. $\varphi^\star+\psi^\star = 0$ will again be the best approximation to f on Q. Hence, the subspace $D(Q)$ is not everywhere dense in $C(Q)$. At the same time, Q does not contain any closed lightning bolts.

An addition to the duality theorem. We can now prove Corollary 4.26, but with an essential extra assumption: existence of the best approximation $\varphi^\star(x)+\psi^\star(y)$ for f. Indeed, in view of Lemma 4.5, it is obvious that the left side in (4.37) is not larger than the right side. On the other hand, if ℓ is the lightning bolt described in Criterion 6.1, then

$$|r_\ell(f)| = M = \|f-\varphi^\star-\psi^\star\| = E(f).$$

The following theorem of de la Vallée–Poussin concerning an estimate from below of the best approximation (cf., e.g., [**2**]) plays a useful role in the theory of the best polynomial approximation. For our problem there is an analogue of that result.

THEOREM 6.10 [**79**]. *Let the difference $f(x,y)-\varphi(x)-\psi(y)$ assume the values a_1, $-a_2$, $a_3,\dots$ at the vertices $p_1,p_2,p_3,\dots$ of a lightning bolt ℓ, where $a_i>0$ (and hence, signs at the neighboring vertices alternate) while ℓ is either closed or infinite. Then,*

$$E(f)\ge \inf a_i. \tag{6.6}$$

PROOF. Set $a=\inf a_i$. By (4.6)–(4.7) and Lemma 4.5 we have

$$|r_\ell(f)| = |r_\ell(f-\varphi-\psi)| \ge \overline{\lim_n}\frac{1}{n}\left[a_1+\dots+a_n\right]\ge a.$$

Now apply the duality theorem.

AN EXAMPLE OF THE CHEBYSHEV CRITERION IN ACTION. Let us give a non-trivial example when Theorem 6.1 does apply.

PROPOSITION 6.11 [**79**]. *Let Q be a rectangle in $\mathbb{R}^2 : Q=[0,a]\times[0,b]$, and $f(x,y)\in C(Q)$. Set*

$$g(x,y)=f(x,y)-f(x,0)-f(0,y)+f(0,0) \tag{6.7}$$

and suppose that for any fixed y_1 and y_2, $0\le y_2<y_1\le b$, we have

$$g\left(x,y_1\right)-g\left(x,y_2\right)\ge 0,\qquad 0\le x\le a, \tag{6.8}$$

and attains its maximum at one and the same value x_0. Then,

$$E(f)=\frac{1}{4}\left[f\left(x_0,b\right)-f\left(x_0,0\right)-f(0,b)+f(0,0)\right]. \tag{6.9}$$

PROOF. Clearly, $E(f)=E(g)$. Moreover, we have

$$g(0,y)=0,\quad g(x,0)=0,\qquad 0\le x\le a,\quad 0\le y\le b.$$

In view of (6.8) this implies that, for all x and y, $g(x,y)\ge 0$. Set

$$A=g\left(x_0,b\right)=\left[f\left(x_0,b\right)-f\left(x_0,0\right)-f(0,b)+f(0,0)\right].$$

Consider the function

$$h(x,y)=g(x,y)-\frac{1}{2}g\left(x_0,y\right)$$

for which $E(h)=E(g)=E(f)$. For all admissible x and $y_1>y_2$, we have

$$\begin{aligned}
h(x,y_1)-h(x,y_2) &= g(x,y_1)-g(x,y_2)-\frac{1}{2}\left[g(x_0,y_1)-g(x_0,y_2)\right] \\
&\le g(x_0,y_1)-g(x_0,y_2)-\frac{1}{2}\left[g(x_0,y_1)-g(x_0,y_2)\right]=\frac{1}{2}\left[g(x_0,y_1)-g(x_0,y_2)\right] \\
&\le \frac{1}{2}\left[g(x_0,b)-g(x_0,0)\right]=\frac{1}{2}g(x_0,b)=\frac{A}{2},
\end{aligned}$$

$$h(x,y_1)-h(x,y_2)\ge -\frac{1}{2}\left[g(x_0,y_1)-g(x_0,y_2)\right]\ge -\frac{A}{2}.$$

So,

$$|h(x,y_1)-h(x,y_2)|\le \frac{A}{2}.$$

Set

$$k(x)=\frac{1}{2}\left[\sup_y h(x,y)+\inf_y h(x,y)\right],$$
$$m(x,y)=h(x,y)-k(x).$$

We have $E(m)=E(h)=E(f)$. For all x we have

$$\sup_y|m(x,y)|\le \frac{1}{2}\left[\sup_y h(x,y)-\inf_y h(x,y)\right]\le \frac{A}{4}.$$

Hence

$$\|m\|\le\frac{A}{4}.$$

On the other hand,

$$\begin{aligned}
&h(0,0)=0, && h(0,y)=-\frac{1}{2}g(x_0,y), && k(0)=-\frac{A}{4}, && m(0,0)=\frac{A}{4},\\
&h(x_0,0)=0, && h(x_0,y)=\frac{1}{2}g(x_0,y), && k(x_0)=\frac{A}{4}, && m(x_0,0)=-\frac{A}{4},
\end{aligned}$$

$$h(0,b)=-\frac{1}{2}g(x_0,b)=-\frac{A}{2},\qquad m(0,b)=-\frac{A}{2}+\frac{A}{4}=-\frac{A}{4},$$
$$h(x_0,b)=\frac{1}{2}g(x_0,b)=\frac{A}{2},\qquad m(x_0,b)=\frac{A}{2}-\frac{A}{4}=\frac{A}{4}.$$

According to Theorem 6.1, the best approximation to $m(x,y)$ is given by the combination $0+0$, with $E(m)=A/4$. The proposition is proved.

Corollary 6.12 [**116**]. *If $f(x,y)$ is continuous together with its first and second partial derivatives in the rectangle Q and $\partial^2 f/\partial x\,\partial y\ge 0$, then*

$$E(f)=\frac{1}{4}\left[f(0,0)-f(0,b)-f(a,0)+f(a,b)\right]. \tag{6.10}$$

PROOF. In this case the function $g(x,y)$ defined by (6.7) equals

$$\iint\limits_{Q_{xy}} \frac{\partial^2 f}{\partial x\, \partial y} dxdy,$$

where Q_{xy} is the rectangle with vertices $(0,0)$, $(x,0)$, $(0,y)$, (x,y). Therefore, $g(x,y)$ is increasing with respect to x and y, and all the assumptions of Proposition 6.11 are satisfied.

Our goal now is to give a new proof of the existence of a function giving the best approximation. For this we must introduce the following elementary operator.

The levelling operator. Consider a space of bounded, real-valued functions $v(x)$ defined on a set $E \subset (-\infty, +\infty)$ with the norm $\|v\| = \sup_{x\in E} |v(x)|$. Introduce the operator (functional) M by

$$Mv = \frac{1}{2}\left[\sup_{x\in E} v(x) + \inf_{x\in E} v(x)\right]. \tag{6.11}$$

For the function $u(x) = v(x) - Mv$ we have $\inf u(x) = -\sup u(x)$, and so we call M the levelling operator.

LEMMA 6.13 [**62**]. *For the operator M the following properties are satisfied:*
1. *Monotonicity:*

$$u(x) \le v(x) \text{ for all } x \in E \Rightarrow Mu \le Mv. \tag{6.12}$$

2.

$$|Mv| \le \|v\|. \tag{6.13}$$

3.

$$|Mv_1 - Mv_2| \le \|v_1 - v_2\|. \tag{6.14}$$

4.

$$\|v - Mv\| = \|v\| - \|Mv\| \le \|v\|. \tag{6.15}$$

Properties 1 and 2 are obvious. To prove (6.14) set $\delta = \|v_1 - v_2\|$. Then, $-\delta \le v_1(x) - v_2(x) \le \delta$, $-\delta + v_2(x) \le v_1(x) \le \delta + v_2(x)$. By monotonicity we obtain

$$M[-\delta + v_2(x)] = -\delta + Mv_2 \le Mv_1 \le M[\delta + v_2(x)] = \delta + Mv_2,$$

and therefore, $-\delta \le Mv_1 - Mv_2 \le \delta$. To prove (6.15), set $\alpha = \sup v(x)$, $\beta = \inf v(x)$. Then $Mv = \frac{1}{2}(\alpha+\beta)$, $\|v - Mv\| = \frac{1}{2}(\alpha - \beta)$, and $\|v\| = \max(\alpha, -\beta)$. However, it is easy to check that always

$$\frac{1}{2}(\alpha - \beta) = \max(\alpha, -\beta) - \frac{1}{2}|\alpha + \beta|.$$

Levelling of functions of two variables. Let X, Y, $Q \subset X \times Y$ be arbitrary sets; π_1, π_2 are natural projections onto X and Y, respectively, and we can assume that $X = \pi_1(Q)$, $Y = \pi_2(Q)$. Similarly to (2.3), (2.18), for a function $f(x,y)$ bounded on Q, set

$$\begin{aligned} M(x) &= \sup_{y,(x,y)\in Q} f(x,y), \qquad & m(x) &= \inf_{y,(x,y)\in Q} f(x,y), \\ M(y) &= \sup_{x,(x,y)\in Q} f(x,y), \qquad & m(y) &= \inf_{x(x,y)\in Q} f(x,y), \end{aligned} \tag{6.16}$$

and define the operators

$$\begin{aligned} M_x(f)(y) &= \frac{1}{2}\left[M(y) + m(y)\right], \\ M_y(f)(x) &= \frac{1}{2}\left[M(x) + m(x)\right]. \end{aligned} \tag{6.17}$$

Then, the function $f(x,y) - M_y(f)(x)$ is levelled with respect to y, while $f(x,y) - M_x(f)(y)$ is levelled with respect to x. The function $M_y(f)(x)$ is a solution of the problem (2.17) concerning the best approximation of $f(x,y)$ by functions of one variable x, and $M_x(f)(y)$ is a solution of a similar problem concerning the best approximation of $f(x,y)$ by functions of y. If X, Y are compact spaces, $Q = X \times Y$, $f \in C(Q)$, then the functions (6.17) solve those problems of best approximation in the space $C(Q)$.

Existence theorem. Let us return to problem (6.1) and give a new proof of existence of the best approximation $\varphi^\star(y) + \psi^\star(y)$ in the case when $Q = X \times Y$. In addition, we shall also obtain an estimate concerning the type of continuity of $\varphi^\star + \psi^\star$ in terms of that of the function being approximated. Fix $f \in C(Q)$ and define (nonlinear) operators in D

$$\begin{aligned} AF &= M_x(f - F), \\ BF &= M_y(f - F). \end{aligned} \tag{6.18}$$

From (6.14) it follows immediately that

$$\begin{aligned} \|AF_1 - AF_2\| &\le \|F_1 - F_2\|, \\ \|BF_1 - BF_2\| &\le \|F_1 - F_2\|. \end{aligned} \tag{6.19}$$

Also, define an operator $S : D \to D$ as follows:

$$SF = F + AF + B(F + AF), \qquad F \in D. \tag{6.20}$$

Introduce the following characteristic of the uniform continuity of the function $f(x,y)$:

$$\Delta(x_1, y_1, x_2, y_2) = \sup_x |f(x, y_1) - f(x, y_2)| + \sup_y |f(x_1, y) - f(x_2, y)| \tag{6.21}$$

and define a set $K \subset D$ by

$$K = \{F \in D : \|F - f\| \le \|f\|,\ |F(x_1, y_1) - F(x_2, y_2)| \le \Delta(x_1, y_1, x_2, y_2)\}. \tag{6.22}$$

THEOREM 6.14 [**91**]. *K is a non-empty, convex, compact set in $C(Q)$. The operator S is continuous and $S(K) \subset K$, so S has fixed points. If F is a fixed point of S in D, then $f - F$ is a levelled function and F is the best approximation of f in D.*

Thus, among functions F giving the best approximation of f in D, there are some for which $f - F$ is a levelled function, and uniform continuity of F satisfies the same estimate as that of f.

PROOF. That K is non-empty is obvious, since $0 \in K$. It is also easy to check that K is convex. The set $\tilde{K}$ of all functions in $C(Q)$ satisfying the inequalities in (6.22) is bounded, closed, and equicontinuous; therefore, it is compact in $C(Q)$. Since $K = \tilde{K} \cap D$ and D (e.g., according to Corollary 2.7) is closed, K is compact. As follows from (6.19), the operators A and B are continuous; hence, S is a continuous operator. Let us show that $S(K) \subset K$. For $F \in D$ we have (applying (6.15) twice)

$$\begin{aligned} \|f - SF\| &= \|f - F - AF - B(F + AF)\| \\ &= \|f - F - AF - M_y(f - F - AF)\| \\ &\leq \|f - F - Af\| = \|f - F - M_x(f - F)\| \leq \|f - F\|. \end{aligned} \tag{6.23}$$

This shows that if $F \in K$, then $\|f - SF\| \leq \|f\|$. If $F = g(x) + h(y)$, then using the definitions of M_x and M_y we find

$$\begin{aligned} SF &= g + h + M_x(f - g - h) + B(F + AF) \\ &= g + h + M_x(f - g) - h + M_y\left[f - F - M_x(F - f)\right] \\ &= g + Ag + M_y\left[f - g - h + M_x(f - g) + h\right] \\ &= g + Ag + M_y\left[f - M_x(f - g)\right] - g = Ag + BAg. \end{aligned} \tag{6.24}$$

Let $F = g(x) + h(y)$ and $\Phi(x, y) = SF$. Using Lemma 6.13 and (6.24), we obtain

$$\begin{aligned} &|\Phi(x_1, y_1) - \Phi(x_2, y_2)| \\ &\quad = |(Ag)(x_1, y_1) - (Ag)(x_2, y_2) + (BAg)(x_1, y_1) - (BAg)(x_2, y_2)| \\ &\quad = |(Ag)(y_1) - (Ag)(y_2) + (BAg)(x_1) - (BAg)(x_2)| \\ &\quad \leq |M_x(f - g)(y_1) - M_x(f - g)(y_2)| \\ &\qquad + |M_y(f - Ag)(x_1) - M_y(f - Ag)(x_2)| \\ &\quad \leq \sup_x |(f - g)(x, y_1) - (f - g)(x, y_2)| \\ &\qquad + \sup_y |(f - Ag)(x_1, y) - (f - Ag)(x_2, y)| \\ &\quad = \sup_x |f(x, y_1) - f(x, y_2)| + \sup_y |f(x_1, y) - f(x_2, y)| \\ &\quad = \Delta(x_1, y_1, x_2, y_2). \end{aligned} \tag{6.25}$$

So, (6.23) and (6.25) show that $S(K) \subset K$. Hence, by the Schauder theorem the operator S has a fixed point in K. Let F be a fixed point of S in D. Then

$$AF + B(F + AF) = 0. \tag{6.26}$$

Since AF is a function of y only while $B(F + AF)$ is a function of x, then $AF = C$, $B(F + AF) = -C$, where C is a constant. Hence, $-C = B(F + C) = BF - C$ and

$BF = 0$. So, $M_y(f - F) = 0$ and the function $f - F$ is vertically levelled. Also, $AF = M_x(f - F) = C$. However, we are going to show that $C = 0$.

LEMMA 6.15 [**91**]. *If* $M_y f = 0$ *while* $M_x f = C$, *then* $C = 0$.

PROOF. Applying Lemma 6.13, we find that

$$\|f\| = \|f - C - M_y f + C\| = \|(f - C) - M_y(f - C)\| \leq \|f - C\|$$
$$= \|f - M_x f\| = \max_y \left[\max_x |f| - |M_x f|\right] = \|f\| - |C| \quad \text{and } C = 0.$$

Lemma 6.15 is proved.

So in the situation described by Theorem 6.14, $M_x(f - F)$ and hence $f - F$ is a levelled function. But then, according to Corollary 6.4, $E(f) = \|f - F\|$ and F is the best approximation to f in D.

REMARKS 6.16. (a) For the first time, existence of a best approximation with the same estimates for continuity as those of the approximated function (defined on the rectangle $[a, b] \times [c, d] \subset \mathbb{R}^2$) was proved in the paper of Diliberto and Straus [**38**] by a different method that was, in a sense, more natural and elementary. (We shall discuss it in the following section.) Independently, Kolmogorov (see [**113**]) obtained the same result by a method close to the one applied in [**38**]. In [**61**] there are some results concerning smoothness of the function giving the best approximation in terms of the smoothness of the approximated function.

(b) In the derivation of estimates (6.23) and (6.25), continuity of the function g and h has not been used. Hence, if g and h are discontinuous functions, then $S(g + h)$ is nevertheless continuous and, in view of (6.23),

$$\|f - S(g + h)\| \leq \|f - (g + h)\|.$$

So, approximation of a continuous function f does not improve if we allow discontinuous g and h. This was noted for the first time in [**32**] by a different reasoning. In §2 of Chapter 3 we shall establish a substantial extension of this fact (Theorem 2.4, Chapter 3).

(c) Consider approximation of $f \in B(Q)$ by a subspace BD (cf. §§2, 3). The following statement is true then: if f has the best approximation $\varphi^\star(x) + \psi^\star(y) \in BD$, then among its best approximations there is one $\varphi_1(x) + \psi_1(y)$ such that $f - (\varphi_1 + \psi_1)$ is a vertically levelled function, and similarly, there is a best approximation $\varphi_2(x) + \psi_2(y)$ so that $f - (\varphi_2 + \phi_2)$ is horizontally levelled. Indeed, if $f^\star = f - (\varphi^\star + \psi^\star)$, then the function $f_1 = f^\star - M_y f^\star$ is vertically levelled with $\|f_1\| \leq \|f^\star\|$, according to Lemma 6.13. Similarly, $f_2 = f^\star - M_x f^\star$ is horizontally levelled with $\|f_2\| \leq \|f^\star\|$.

Example of deterioration of continuity properties for the best approximation. In the case when $Q \neq X \times Y$, given the existence of the function of the best approximation, its continuity properties may be worse than those of the function being approximated.

EXAMPLE 6.17 [**113**]. $Q = Q_1 \cup Q_2 \cup Q_3$; $Q_1 = \{(x, 1),\ 1 \leq x \leq 2\}$, $Q_2 = \{(x, 2),\ 1 \leq x \leq 3\}$, $Q_3 = \{(x, 3),\ 2 + \varepsilon \leq x \leq 3\}$, where $\varepsilon > 0$ is given. Assume $f(x, y) = 0$ on Q_2 while on Q_1 and Q_3 the graph of $f(x, y)$ is as in Figure 1 (the plane of the drawings coincide with XOZ). Let the graph of $\psi_0(x)$ be shown

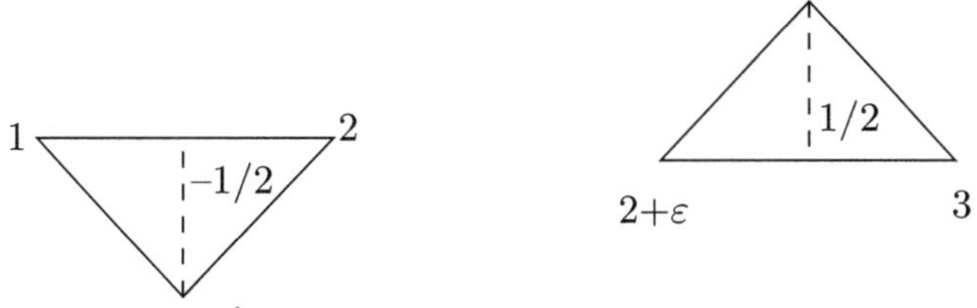

FIGURE 1

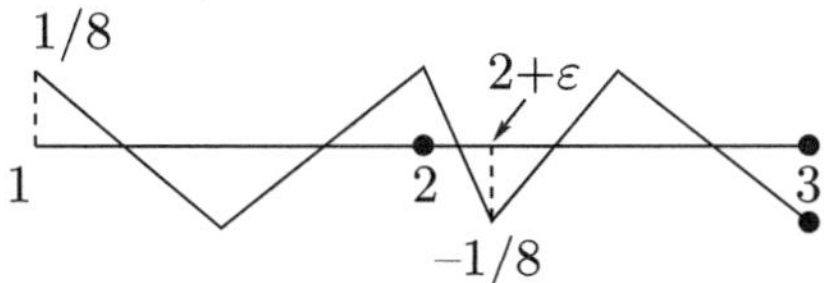

FIGURE 2

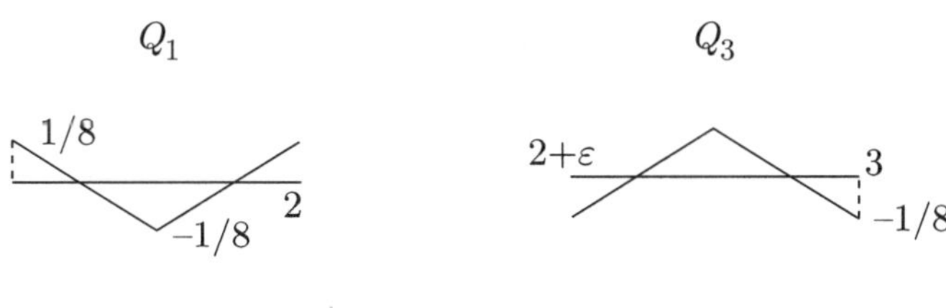

for Q_2: $f_1(x, 2) = -\varphi_0(x)$

FIGURE 3

in Figure 2 and $\psi_0(1) = -\frac{1}{4}$, $\psi_0(2) = 0$, $\psi_0(3) = \frac{1}{4}$. The graph of $f_1(x,y) = f(x,y) - \varphi_0(x) - \psi_0(y)$ is shown in Figure 3. Theorem 6.1 or Corollary 6.4 yields that $\varphi_0(x) + \psi_0(y)$ is the best approximation and $E(f) = \frac{1}{8}$. Also, it is easy to see that $\psi_0(y)$ is a unique (up to a constant) function that can be included into the best approximation. Regarding $\varphi^\star(x)$, it can differ from $\varphi_0(x)$, but since $f(2,1) - \psi_0(1) = \frac{1}{4}$, $f(2+\varepsilon,3) - \psi_0(3) = -\frac{1}{4}$ and $E(f) = \frac{1}{8}$, then necessarily $\varphi^\star(2) \geq \frac{1}{8}$, $\varphi^\star(2+\varepsilon) \leq -\frac{1}{8}$. The function $f(x,y)$ satisfies a Lipschitz condition with a constant L_0 that does not depend on ε. At the same time, for any $L > L_0$ we can choose $\varepsilon > 0$ sufficiently small so that $\varphi^\star(x)$ does not satisfy the Lipschitz condition with that constant L.

In [**113**], Ofman claimed (Theorem 3) that if Q contains a cross while $f(x,y)$ satisfies a Lipschitz condition with a constant L, then f has a best approximation in D also satisfying the Lipschitz condition with the same constant L. Yet, as was noted by Motornyi [**109**], this is false. In order to see that, it suffices to use the original example of Ofman.

EXAMPLE 6.18 [**105**]. Add a point $(3,1)$ to Q from Example 6.17. We obtain a compact $\tilde{Q}$. $\tilde{Q}$ contains a cross passing through the point $(3,2)$. Define the function f at the point $(3,1)$ by setting $f(3,1) = -\frac{1}{4}$. It is easy to see that the same $\varphi^\star(x) + \psi_0(y)$ as in Example 6.17 give the best approximation here too.

In [**109**], [**110**] there were given sufficient conditions for Q in order that the best approximation in D of $f \in C(Q)$ had the same majorant of uniform continuity. However, the arguments presented there contain a gap.

Best approximation and diameters of families of functions of one variable. Let us present a somewhat different interpretation of the problem of the best approximation that we have been studying. Let X be an arbitrary set, and let $V = \{g(x)\}$ be a given family of real-valued functions (not necessarily continuous) of one variable. Each function $g(x) \in V$ has, generally speaking, its own domain of definition $A_g \subset X$, and it is not necessarily true that $A_{g_1} = A_{g_2}$. Assume that functions in V are uniformly bounded. Define a distance between functions in the family by setting

$$\rho\left(g_1, g_2\right) = \sup_{x \in A_{g_1} \cap A_{g_2}} \left|g_1(x) - g_2(x)\right|. \tag{6.27}$$

If $A_{g_1} \cap A_{g_2} = \emptyset$, set $\rho\left(g_1, g_2\right) = 0$. (In general, this "distance" does not satisfy the axioms of a metric space.) Naturally, we shall call the quantity

$$d(V) = \sup_{g_1, g_2 \in V} \rho_V\left(g_1, g_2\right) \tag{6.28}$$

the diameter of V. Let us add to each function $g(x)$ of the family V a constant C_g, one for each g. We denote the new family obtained in such a way by $V + C$ (C is the family of constants added). The quantity

$$d_0(V) = \inf_C d(V + C) \tag{6.29}$$

is called the proper diameter of V (here the inf is taken over all families of added constants). Let $f(x, y)$ be a bounded function defined on a set $Q \subset X \times Y$, where X,Y are arbitrary sets. Consider families $V_1 = \{g_y(x) = f(x, y),\ \ y \in Y\}$ and $V_2 = \{h_x(y) = f(x, y), x \in X\}$. Let d_0^i be the proper diameter of V_i, $i = 1, 2$.

THEOREM 6.19 [**113**]. *The following equalities hold:*

$$\mathcal{E}(f) \stackrel{\text{def}}{=} \operatorname{dist}(f, BD) = \frac{1}{2} d_0^1 = \frac{1}{2} d_0^2. \tag{6.30}$$

PROOF. First of all, note that for a function $g_{y_0}(x) \in V_1$ the domain of definition is $A_y = \pi_1\left\{(x, y_0) \in Q\right\}$. Let $\psi(y)$ be an arbitrary bounded function on Y. We shall interpret it as a family of constants that are added to functions in the family V_1. Introduce the following notation (x is fixed):

$$\begin{aligned} M_\psi(x) &= \sup_{g_y \in V_1,\, A_y \ni x} \left[g_y(x) - \psi(y)\right] = \sup_{A_y \ni x} \left[f(x, y) - \psi(y)\right]; \\ m_\psi(x) &= \inf_{g_y \in V_1,\, A_y \ni x} \left[g_y(x) - \psi(y)\right] = \inf_{A_y \ni x} \left[f(x, y) - \psi(y)\right]; \\ \varphi_\psi(x) &= \frac{M_\psi(x) + m_\psi(x)}{2}. \end{aligned} \tag{6.31}$$

Clearly,

$$d\left(V_1 - \psi\right) = \sup_{x \in X} \left[M_\psi(x) - m_\psi(x)\right]. \tag{6.32}$$

On the other hand, for any $\varphi(x)$ defined on X we have
(6.33)

$$\|f-\varphi-\psi\| \geq \|f-\varphi_\psi-\psi\| = \frac{1}{2}\sup_{x\in X}\left[M_\psi(x)-m_\psi(x)\right] = \frac{1}{2}d\left(V_1-\psi\right) \geq \frac{1}{2}d_0^1.$$

Hence,

$$\mathcal{E}(f) \geq \frac{1}{2}d_0^1. \tag{6.34}$$

On the other hand, let $\varepsilon > 0$ and $\psi_0(y)$ be such that

$$d\left(V_1-\psi_0\right) < d_0^1+\varepsilon. \tag{6.35}$$

Then, in view of (6.33),

$$\|f-\varphi_{\psi_0}-\psi_0\| = \frac{1}{2}d\left(V_1-\psi_0\right) < \frac{1}{2}d_0^1+\frac{\varepsilon}{2},$$

and so,

$$\mathcal{E}(f) \leq \frac{1}{2}d_0^1. \tag{6.36}$$

Combining (6.34) and (6.35) we obtain the first equality in (6.30). The proof of the second one is similar.

Corollary 6.20. *Let $Q = X\times Y$, X and Y compact spaces, and $f\in C(Q)$. Then*

$$E(f) = \frac{1}{2}d_0^1 = \frac{1}{2}d_0^2. \tag{6.37}$$

For the proof, it suffices to combine Theorem 6.19 and Remark 6.16 (b).

Set of functions giving the best approximation. Usually, a function giving the best approximation in the class D is not unique. Let $f(x,y)\in C(Q)$, and let $f^\star(x,y) = f(x,y)-\varphi^\star(x)-\psi^\star(y)$ be a levelled function. Since in some sense $\varphi^\star(x)+\psi^\star(y)$ gives the "best" best approximation to $f(x,y)$, it is natural to ask whether such a function $\varphi^\star(x)+\psi^\star(y)$ is unique. The following example shows that even in such a form uniqueness, generally speaking, fails.

Example 6.21. Let $Q = X\times Y$, where X and Y consist of four points each. In this case, $f(x,y)$ is a square matrix. Consider

$$f = \begin{pmatrix} 1 & 0 & -1 & 0 \\ 0 & -1 & 0 & 1 \\ 0 & 1 & 0 & -1 \\ -1 & 0 & 1 & 0 \end{pmatrix}.$$

Clearly, f is a levelled function, and hence $E(f) = \|f\| = 1$. Take

$$\varphi^\star = (\varepsilon, 2\varepsilon, \varepsilon, 2\varepsilon), \qquad \psi^\star = \begin{pmatrix} -\varepsilon \\ -2\varepsilon \\ -2\varepsilon \\ \varepsilon \end{pmatrix},$$

(ε is an arbitrary number, $0<\varepsilon<\frac{1}{2}$). Then $f^\star = f-\varphi^\star-\psi^\star$ is also a levelled function and $E\left(f^\star\right) = \|f^\star\| = 1$.

Now we present a number of elegant results, most of which are taken from [**61**] and [**91**] and concern certain sets that appear naturally in Chebyshev's approximation by a subspace D.

Clearly, Example 6.21 can be easily extended to the case when Q is a rectangle. Let $Q \subset X \times Y$, Q, X, and Y compact sets, and $f \in C(Q)$. Denote by $\mathcal{M}$ the set of all signed measures $\mu^\star \in D(Q)^\perp$ for which the supremum in (4.11) is attained and $\int_Q f d\mu^\star > 0$. Set

$$S(f) = S = \bigcup_{\mu^\star \in \mathcal{M}} S(\mu^\star). \tag{6.38}$$

THEOREM 6.22. *All best approximations $\varphi^\star(x) + \psi^\star(y)$ to f coincide on the set S.*

The proof follows from Theorem 3.2 (formula (3.2)). Denote by $Z = Z(f)$ the set of points in Q where all functions $\varphi^\star(x) + \psi^\star(y)$ giving the best approximation to f coincide. Theorem 6.22 claims that

$$Z(f) \supset S(f). \tag{6.39}$$

We can assume that $E(f) = \|f\|$, i.e., the identically zero function is one of the best approximations to f (for that, one may need to pass from f to $f - \varphi^\star - \psi^\star$, where $\varphi^\star(x) + \psi^\star(y)$ is one of the best approximations). Then

$$Z = \bigcap_{\Phi} Z_\Phi, \tag{6.40}$$

where Z_Φ is the set of zeros of the function $\Phi = \varphi^\star(x) + \psi^\star(y)$ and the intersection is taken over all Φ that give the best approximation to f. Together with S and Z, consider also the set

$$N_f = \{p \in Q : |f(p)| = \|f\|\} \tag{6.41}$$

(we assume that $E(f) = \|f\|$). From Theorem 6.2 it follows that

$$N_f \supset S. \tag{6.42}$$

Denote by $\mathcal{P}$ the following set of functions:

$$\mathcal{P} = \mathcal{P}(f) = \{f_1 : f_1 - f \in D, \|f_1\| = \|f\|(= E(f))\}. \tag{6.43}$$

Consider the set N given by

$$N = N(f) = \bigcap N_{f_1} \supseteq S, \tag{6.44}$$

where the intersection is taken over all functions $f_1 \in \mathcal{P}$. The sets Z and N are compact.

LEMMA 6.23. *If there is a function $f^\star \in \mathcal{P}$ for which*

$$N_{f^\star} = N(f), \tag{6.45}$$

then

$$Z(f) \supseteq N(f). \tag{6.46}$$

Thus,

$$Z(f) \supseteq N(f) \supset S(f). \tag{6.47}$$

PROOF. Let $f_1 \in \mathcal{P}(f)$. Then, $F = \frac{1}{2}(f + f_1)$ also belongs to $\mathcal{P}(f)$. Clearly,

$$N_F \subseteq N_{f^\star} \cap N_{f_1} \subseteq N_{f^\star} = N.$$

Therefore, $N_F = N$. If at a point $p \in N$ we had

$$f_1(p) \neq f^\star(p) \qquad (|f^\star(p)| = \|f^\star\| = \|f_1\|),$$

then $p \notin N_F$ and $N_F \neq N$. Hence all functions $f_1 \in \mathcal{P}$ coincide with $f^\star$ on N, and so best approximation functions to f (precisely those that distinguish between different functions in $\mathcal{P}$) equal zero.

Existence of a function $f^\star \in \mathcal{P}$ satisfying the properties listed in Lemma 6.23 probably need not hold in general. However, if we restrict ourselves to, say, functions satisfying inequalities (6.22) and change the definitions of $\mathcal{P}$ and N accordingly, then among such functions a function $f^\star$ satisfying the requirements of Lemma 6.23 exists in view of compactness of the set K.

A geometric lemma.

LEMMA 6.24 [**91**]. *Let R be a subset of $X \times Y$, where X and Y are arbitrary sets having the following properties*:

$$\{(x, y) \in R,\ (u, y) \in R,\ (u, v) \in R\} \Rightarrow (x, v) \in R. \tag{6.48}$$

Then there exist a family of pairwise disjoint sets $\{X_\alpha\} \subset X$ and a family of pairwise disjoint sets $\{Y_\alpha\} \subset Y$ such that $R = \bigcup_\alpha X_\alpha \times Y_\alpha$.

Geometrically, (6.48) means that when three vertices of a rectangle with sides parallel to the coordinate axes belong to R, the fourth vertex also belongs to R.

PROOF. Set

$$X_0 = \{x \in X : (x, y) \notin R \quad \text{for all } y \in Y\},$$
$$Y_0 = \{y \in Y : (x, y) \notin R \quad \text{for all } x \in X\}.$$

On $X \backslash X_0$ we define an equivalence relation by setting $x_1 \sim x_2$ if and only if there exists $y \in Y$ such that $(x_1, y) \in R$ and $(x_2, y) \in R$. Using (6.48) one can easily check that this is indeed an equivalence relation. Let $\{X_\alpha\}$ be the family of equivalence classes defined by this relation. Define

$$Y_\alpha = \{y \in Y : (x, y) \in R \quad \text{for all } x \in X_\alpha\}.$$

Let us show that different sets Y_α do not intersect. Assume that $y \in Y_\alpha \cap Y_\beta$. Then $(x, y) \in R$ for all $x \in X_\alpha \cap X_\beta$. Choose $x_\alpha \in X_\alpha$ and $x_\beta \in X_\beta$. Then $(x_\alpha, y) \in R$ and $(x_\beta, y) \in R$, i.e., $x_\alpha \sim x_\beta$ and therefore $X_\alpha = X_\beta$. Let us show that $R \supset \bigcup_\alpha X_\alpha \times Y_\alpha$. Let $(u, v) \in X_\alpha \times Y_\alpha$. By construction of Y_α, $(x, v) \in R$ for all $x \in X_\alpha$ and, in particular, for $x = u$, i.e., $(u, v) \in R$. It remains to show that $R \supset \bigcup_\alpha X_\alpha \times Y_\alpha$.

Let $(a,v)\in R$. Then $u\in X\backslash X_0$, and therefore $u\in X_\alpha$ for some α. Let x be an arbitrary element in X_α. Then $u\sim x$, and so there exists $y\in Y$ such that $(x,y)\in R$ and $(u,y)\in R$. Now we have $\{(u,y)\in R,\ (x,y)\in R,\ (u,v)\in R\}\Rightarrow(x,v)\in R$ in view of (6.48). So, not only $(u,v)\in R$ but also $(x,v)\in R$ for all $x\in X_\alpha$. Hence, $v\in Y_\alpha$ and $(u,v)\in X_\alpha\times Y_\alpha$.

LEMMA 6.25. *Let $R\subset X\times Y$ and satisfy (6.48). Let $p_0\in(X\times Y)\backslash R$. Then there exists a function $\Phi(x,y)=g(x)+h(y)$ such that $\Phi=0$ on R and $\Phi(p_0)=1$.*

PROOF. Let X_0, $\{X_\alpha\}$, Y_0, $\{Y_\alpha\}$ be as in Lemma 6.24. Then

$$X=X_0\cup\bigcup_\alpha X_\alpha,\qquad Y=Y_0\cup\bigcup_\alpha Y_\alpha,\qquad R=\bigcup_\alpha X_\alpha\times Y_\alpha.$$

Let $p_0=(x_0,y_0)$. If $x_0\in X_0$, then set $g(x_0)=1$ and $g(x)=0$ elsewhere, and let $h(y)\equiv0$. Then $g+h$ is the needed function. If $x_0\notin X_0$, then $x_\alpha\in X_\alpha$ for some α. Set $g(x)=1$, $x\in X_\alpha$, $g(x)=0$ elsewhere; $h(y)=-1$, $y\in Y_\alpha$, $h(y)$ elsewhere on Y. The function $g+h$ satisfies the required properties.

The set on which all best approximations coincide.

THEOREM 6.26 [**91**]. *Let $Q=X\times Y$, where X and Y are finite sets and f is a given function on Q. The set of points $Z(f)$ where all best approximations to f by the functions $\varphi(x)+\psi(y)$ coincide is the smallest set satisfying (6.48) and containing the set $N(f)$ of vertices of all extremal lightning bolts.*

PROOF. Without loss of generality we can assume that $N(f)=N_f$. According to Lemma 6.23, $Z(f)\supset N(f)$. If at three vertices of a rectangle with sides parallel to the coordinate axes one of the best approximations $\Phi=\varphi(x)+\psi(y)$ to f vanishes, then it also vanishes at the fourth vertex. Indeed, if ℓ is the lightning bolt defined by those vertices, then $r_\ell\in D^\perp$ and $r_\ell(\Phi)=0$. So, (6.48) holds for $D(f)$. Now, let R be a set such that $R\supset N(f)$ and R satisfies (6.48). Let $p_0\notin R$. Then, according to Lemma 6.25, there exists a function $\Phi=g(x)+h(y)$ such that $\Phi=0$ on R and $\Phi(p_0)=1$. It is not difficult to see that since $\Phi=0$ on $N(f)$, the function $\varepsilon\Phi$ has the property $\|f-\varepsilon\Phi\|=\|f\|$ for sufficiently small $\varepsilon>0$ and hence provides the best approximation to f. Therefore, $p_0\notin Z(f)$ since $\varepsilon\phi(p_0)\neq0$. Thus, $R\supseteq Z(f)$, and the theorem is proved.

THEOREM 6.27. *If X and Y are finite sets and $Q=X\times Y$, then for any function f on Q*

$$S(f)=N(f).$$

We shall omit the proof.

Dimension of the set of best approximations.

LEMMA 6.28. *Let a set $R\subseteq X\times Y$ satisfy the property (6.48). If the sets X_0, Y_0 defined in Lemma 6.24 are finite and there exist only a finite number of equivalence classes X_α, then*

$$d\overset{\text{def}}{=}\dim\{\Phi\in D:\ \Phi|_R=0\}=m+n+k-1,\tag{6.49}$$

where m is the number of elements in X_0, n is the number of elements in Y_0, k is the number of classes X_α (dim, *of course, means dimension*).

PROOF. Choose an arbitrary set of k numbers $\beta_1, \ldots, \beta_k$ and define $g(x) = -h(y) = \beta_\alpha$ if $(x, y) \in X_\alpha \times Y_\alpha$, $\alpha = 1, \ldots, k$. Let g and h be arbitrarily defined on X_0 and Y_0. Set $\Phi = g(x) + h(y)$. Clearly, $\Phi \in D$. If $(x, y) \in X_\alpha \times Y_\alpha$, then $\Phi(x, y) = \beta_\alpha - \beta_\alpha = 0$ and so $\Phi|_R = 0$. It is clear from the construction that $d \geq m + n + k - 1$ (the one is subtracted because adding a constant to g and subtracting it from h does not change Φ). On the other hand, let $\Phi = g(x) + h(y)$ and $\Phi|_R = 0$. Then, $g(x) + h(y) = 0$ when $(x, y) \in X_\alpha \times Y_\alpha$. Therefore, $g(x)$ on X_α is equal to a constant β_α, and $h(y) = -\beta_\alpha$ on Y_α. Thus, $d \leq m + n + k - 1$, and the equality (6.49) is proved.

THEOREM 6.29 [**91**]. *Let f be a function on $Q = X \times Y$, where X and Y are finite sets. Let $X = X_0 \cup X_1 \cup \cdots \cup X_k$ and $Y = Y_0 \cup Y_1 \cup \cdots \cup Y_k$ be the decompositions of X and Y defined in Lemma* 6.24 *in connection with $R = Z(f)$. Then*

$$\dim \{\Phi \in D : \|f + \Phi\| = E(f)\} = m + n + k - 1, \tag{6.50}$$

where m is the number of elements in X_0 while n is the number of elements in Y_0.

PROOF. Suppose f is such that $N_f = N(f)$. Each function Φ that gives the best approximation to f must vanish on $Z(f)$. On the other hand, if $\Phi \in D$ is a function such that $\Phi|_{Z(f)} = 0$, then for $\varepsilon > 0$ sufficiently small $\varepsilon\Phi$ is a function of the best approximation to f. It remains to refer to Lemma 6.28 while setting $R = Z(f)$.

Uniqueness of the best approximation.

THEOREM 6.30. *Let X and Y be finite sets and f be a function defined on $Q = X \times Y$. The function f has a unique best approximation in the class D provided that $Z = Q$ or, in other words, if the smallest of the sets that contain $N(f)$ and satisfy* (6.48) *is all of Q.*

PROOF. Uniqueness of the best approximation means that the dimension of a set defined by (6.48) equals zero. This implies that $m = n = 0$ and $k = 1$, i.e., $Z(f) = X \times Y$. Indeed, $k \geq 1$, $m \geq 0$, $n \geq 0$ always.

Lightning bolts and an upper estimate of best approximation. Until now closed lightning bolts have provided a lower estimate of the best approximation. However, they can also be used to obtain estimates of the best approximation from above.

LEMMA 6.31 [**61**]. *Let $Q = X \times Y$, $f(x, y) = f_0(x, y) + \psi(y)$, where f_0 is a vertically levelled function (f, f_0, ψ are in $B(Q)$). Then for any x_1, x_2 in X*

$$\sup_{y \in Y} [f(x_1, y) - f(x_2, y)] \geq 0. \tag{6.51}$$

PROOF. Assume the opposite, i.e., that there exist x_1 and x_2 such that $f(x_1, y) < f(x_2, y)$ for all y. But then $f_0(x_1, y) < f_0(x_2, y)$ for all y, also. Therefore, $M_y f_0(x_1, y) < M_y f_0(x_2, y)$, contrary to f_0 being a vertically levelled function, implying $M_y [f_0(x, y)] = 0$ for all x.

THEOREM 6.32 [**61**]. *Let X, Y be compact spaces, $f \in C(X \times Y)$. Then*

$$E(f) \le \frac{m}{m-1} \sup_{\ell_{2m}} |r_{\ell_{2m}}(f)|, \tag{6.52}$$

where the supremum is taken over all closed lightning bolts ℓ_{2m} that have no more than $2m$ points.

PROOF. According to Theorem 6.14, we can assume f to be a levelled function. Then $E(f) = \|f\|$. Let $\ell = [p_0, p_1, \dots]$, $f(p_i) = (-1)^i \|f\|$, be the lightning bolt mentioned in Theorem 3.1. If ℓ has $2m$ vertices, then (6.52) is obvious, since $r_\ell = \|f\|$. If the number is larger than $2m$ we can take the vertices $p_0, p_1, \dots, p_{2m-3}$ and add to that lightning bolt vertices q_{2m-2} and q_{2m-1} so that $\ell_{2m} = [p_0, \dots, p_{2m-3}, q_{2m-2}, q_{2m-1}]$ is closed and also $f(q_{2m-2}) - f(q_{2m-1}) \ge 0$. The latter is possible in view of Lemma 6.31 (using the fact that f is levelled vertically and horizontally). We have

$$\begin{aligned} r_{\ell_{2m}}(\Phi) &= \frac{1}{2m}\left[\sum_{i=0}^{2m-3} (-1)^i \|f\| + f(q_{2m-2}) - f(q_{2m-1})\right] \\ &\ge \frac{2m-2}{2m}\|f\| = \frac{m-1}{m} E(f). \end{aligned}$$

§7. The levelling algorithm

In this section we study a natural algorithm for constructing the best approximation to a function f by the function $\varphi(x) + \psi(y)$.

The levelling algorithm of Diliberto and Straus ([138]). Assume for now that $Q = X \times Y$, X and Y are compact spaces and $f(x,y) \in C(Q)$. Set

$$\begin{aligned} f_0(x,y) &= f(x,y), \\ g_n(x) &= \frac{1}{2}\left[\max_y f_{n-1}(x,y) + \min_y f_{n-1}(x,y)\right] = M_y(f_{n-1})(x), \\ f_n(x,y) &= f_{n-1}(x,y) - g_n(x) \quad \text{if } n \text{ is odd}, \\ h_n(y) &= \frac{1}{2}\left[\max_x f_{n-1}(x,y) + \min_x f_{n-1}(x,y)\right] = M_x(f_{n-1})(y), \\ f_n(x,y) &= f_{n-1}(x,y) - h_n(y) \quad \text{if } n \text{ is even}. \end{aligned} \tag{7.1}$$

Thus, in accordance with Corollary 1.5, at the first step of the algorithm we construct the best approximation $g_1(x)$ to the function $f(x,y)$ by functions $\varphi(x)$. At the second step we construct the best approximation $h_2(y)$ of the difference $f_1(x,y) = f(x,y) - g_1(x)$ by functions $\psi(y)$; at the third, the best approximation $g_3(x)$ of the function $f_2(x,y) = f_1(x,y) - h_2(y)$ by functions $\varphi(x)$, etc. Now, $f_1(x,y)$ is a vertically levelled function, $f_2(x,y)$ is horizontally levelled, and in general, $f_n(x,y)$ is vertically levelled if n is odd and horizontally levelled if n is even. Clearly,

$$f_n(x,y) = f(x,y) - G_n(x) - H_n(y), \tag{7.2}$$

where

$$(7.3)\qquad \begin{aligned} G_n(x) &= g_1(x) + g_3(x) + \cdots + g_n(x), \\ H_n(y) &= h_2(y) + h_4(y) + \cdots + h_{n-1}(y), \quad \text{if } n \text{ is odd;} \\ G_n(x) &= g_1(x) + \cdots + g_{n-1}(x), \\ H_n(y) &= h_2(y) + \cdots + h_n,(y), \quad \text{if } n \text{ is even.} \end{aligned}$$

We show that when we alternate between the best approximations by functions depending either on x or y, this process leads to the solution of the problem of best approximation by the sums $\varphi(x) + \psi(y)$. Set

$$(7.4)\qquad M_n = \max_{x,y} |f_n(x,y)| = \|f_n\| .$$

Clearly,

$$(7.5)\qquad M_0 \geq M_1 \geq \cdots \geq M_n \geq \cdots \geq E(f),$$

where, as above, $E(f)$ is the distance from f to the functions $\varphi(x) + \psi(y) \in D$. Hence, there exists the limit

$$(7.6)\qquad M = \lim_{n\to\infty} M_n \geq E(f).$$

THEOREM 7.1. *We have*

$$(7.7)\qquad \lim_{n\to\infty} M_n = E(f).$$

PROOF. Following [**38**], let us prove both (7.7) and the formula

$$(7.8)\qquad \sup_{\ell\subset Q} |r_\ell(f)| = E(f),$$

where the supremum is taken over all closed lightning bolts ℓ. Formula (7.8) is, of course, familiar to us—it is the duality theorem proved earlier (even in a more general form) in Theorem 4.8, formula (4.14). Thus, the duality formula will get a new proof here.

On one hand, since $r_\ell \in D^\perp$, we have

$$(7.9)\qquad \sup_{\ell\subset Q} |r_\ell(f)| = \sup_{\ell\subset Q} |r_\ell(f - \varphi - \psi)| \leq \inf_{\varphi+\psi\in D} \|f - \varphi - \psi\| = E(f) \leq M.$$

On the other hand, we shall show that for each $\varepsilon > 0$ we can choose a closed lightning bolt ℓ such that

$$(7.10)\qquad \|r_\ell(f)\| \geq M - \varepsilon.$$

Formulas (7.7) and (7.8) immediately follow from (7.9) and (7.10). Thus, let $\varepsilon > 0$ be arbitrary. Choose a natural number M and a number $\varepsilon_1 > 0$ so that

$$(7.11)\qquad \frac{\|f\|}{2m+2} < \frac{\varepsilon}{4}, \qquad \frac{2m+1}{2m+2} M > M - \frac{\varepsilon}{2}, \qquad 2^{2m}\varepsilon_1 = \frac{\varepsilon}{4}.$$

There exists a number N such that for $n \geq N$

$$(7.12)\qquad M \leq M_n \leq M + \varepsilon_1.$$

To fix the ideas, assume that N is odd and analyze the levelling process at the stages $N, N+1, \ldots, N+2m$, going from the larger indices to the smaller ones.

Since $N+2m$ is odd together with N, f_{N+2m} is a vertically levelled function. Therefore, there exist points (x_1,y_1) and (x_1,y_2) such that

$$f_{N+2m}(x_1,y_1)=M_{N+2m}, \qquad f_{N+2m}(x_1,y_2)=-M_{N+2m}.$$

This means that

$$\begin{aligned} f_{N+2m-1}(x_1,y_1)-g_{N+2m}(x_1)&=M_{N+2m};\\ f_{N+2m-1}(x_1,y_2)-g_{N+2m}(x_1)&=-M_{N+2m}. \end{aligned} \tag{7.13}$$

Therefore,

(7.14)

$$\begin{aligned} g_{N+2m}(x_1)&=f_{N+2m-1}(x_1,y_1)-M_{N+2m}\le M_{N+2m-1}-M_{N+2m}<\varepsilon_1;\\ g_{N+2m}(x_1)&=f_{N+2m-1}(x_1,y_2)+M_{N+2M}\ge-(M_{N+2m-1}-M_{N+2m})>-\varepsilon_1. \end{aligned}$$

From (7.14) we have

$$\begin{aligned} f_{N+2m-1}(x_1,y_1)&>M-\varepsilon_1;\\ f_{N+2m-1}(x_1,y_2)&<-M+\varepsilon_1. \end{aligned} \tag{7.15}$$

Since $f_{N+2m-1}(x,y)$ is a horizontally levelled function, there exist points (x_2,y_2) and (x_2',y_1) such that

$$\begin{aligned} f_{N+2m-1}(x_2,y_2)&>M-\varepsilon_1;\\ f_{N+2m-1}(x_2',y_1)&<-M+\varepsilon_1. \end{aligned} \tag{7.16}$$

We have

$$f_{N+2m-1}(x,y)=f_{N+2m-2}(x,y)-h_{N+2m-1}(y).$$

Substitute this into (7.15) and (7.16), and then replace the values of f_{N+2m-2} at (x_1,y_1) and (x_2,y_2) by $M+\varepsilon_1$, and at (x_1,y_2) and (x_2',y_1) by $-M-\varepsilon_1$ (this will only strengthen the inequalities (7.15) and (7.16)). From the new inequality we obtain the following estimate:

$$\begin{aligned} -2\varepsilon_1<h_{N+2m-1}(y_1)&<2\varepsilon_1;\\ -2\varepsilon_1<h_{N+2m-1}(y_2)&<2\varepsilon_1. \end{aligned} \tag{7.17}$$

Now come back to the inequalities (7.15) and (7.16), where again we set $f_{N+2m-1}=f_{N+2m-2}-h_{N+2m-1}$. We obtain:

$$\begin{aligned} f_{N+2m-2}(x_1,y_1)&>M-3\varepsilon_1;\\ f_{N+2m-2}(x_1,y_2)&<-M+3\varepsilon_1;\\ f_{N+2m-2}(x_2,y_2)&>M-3\varepsilon_1. \end{aligned} \tag{7.18}$$

(A similar inequality for (x_2',y_1) will not be needed.) Since f_{N+2m-2} is a vertically levelled function, there exists a point (x_2,y_3) such that

$$f_{N+2m-2}(x_2,y_3)<-M+3\varepsilon_1. \tag{7.19}$$

Again, take into account that $f_{N+2m-2}(x,y)=f_{N+2m-3}(x,y)-g_{N+2m-3}(x)$, etc. As a result of this "backward" motion from f_{N+2m} to f_N we obtain the points ($2m+1$ of them)

$$(x_1,y_1),(x_1,y_2),(x_2,y_3),\dots,(x_m,y_{m+1}),(x_{m+1},y_{m+1}). \tag{7.20}$$

Note that in each step of such "backward" motion at which functions $h_k(y)$ appear in addition to the point listed in (7.20), there also appears one more point (similar to the point (x_2', y_1) above) that is only needed for the estimate of the function $h_k(y)$ but is not included in (7.20). At those points inequalities similar to (7.16), (7.18) will hold:

$$\text{(7.21)}\qquad \begin{aligned} f_N(x_S, y_S) &> M - \left(2^{2(m-1)} - 1\right)\varepsilon_1 > M - 2^m\varepsilon_1 > M - \frac{\varepsilon}{4};\\ f_N(x_{S-1}, y_S) &< -M + \left(2^{2(m-1)} - 1\right)\varepsilon_1 < -M + 2^m\varepsilon_1 < -M + \frac{\varepsilon}{4}. \end{aligned}$$

Add the point (x_{m+1}, y_1) to the lightning bolt (7.20). We obtain a closed lightning bolt ℓ. At the point (x_{m+1}, y_1) we have a trivial estimate:

$$\text{(7.22)}\qquad |f_N(x_{m+1}, y_1)| \le \|f_N\| \le \|f\|.$$

On the closed lightning bolt ℓ construct the functional r_ℓ (the number of vertices is equal to $2m+2$). Using the inequalities (7.21) and (7.22), and the choice of m, we obtain

$$|r_\ell(f)| = |r_\ell(f_N)| > \frac{2m+1}{2m+2}\left(M - \frac{\varepsilon}{4}\right) - \frac{\|f\|}{2m+2} \ge M - \varepsilon.$$

This inequality completes the proof.

Levelled functions and the algorithm.

COROLLARY 7.2. *If $f(x,y)$ is a levelled function, then $E(f) = \|f\|$.*

PROOF. For this case the Diliberto-Straus algorithm gives $f_0 = f_1 = \cdots = f$, and in view of Theorem 7.1 $E(f) = \|f\|$.

Corollary 7.2 repeats Corollary 6.4, obtained by a different argument.

Before passing to estimates associated with the levelling algorithm let us give an instructive example. If f is a levelled function, the levelling algorithm is not running and there is no need for it. However, if f is not levelled, the algorithm still may not diminish its norm after the first several steps.

EXAMPLE 7.3 [**61**]. For any natural number k there exists a function f such that the functions $\{f_n\}$ in the levelling algorithm satisfy

$$\|f\| = \|f_1\| = \cdots = \|f_{k-1}\| > \|f_k\|.$$

Let $Q = [a,b] \times [c,d] \subset \mathbb{R}^2$. Take the lightning bolt $[p_1, p_2, \dots, p_{2k}]$ and define a function f on the rectangle Q as follows: $f(p_i) = (-1)^i$, $i = 1, \dots, 2k$, and $|f(p)| < 1$ at all remaining points. The function f_1 is vertically levelled, so $|f_1(p_1)| < 1$ and $|f_1(p_{2k})| < 1$, yet $|f_1(p_i)| = 1$, $i = 2, \dots, 2k-1$. The function f_2 is horizontally levelled, so $|f_2(p_i)| < 1$, $i = 1, 2, \dots, 2k-1, 2k$, and $|f_2(p_i)| = 1$, $i = 3, \dots, 2k-2$, etc. Clearly, $|f_k(p)| < 1$ at all points of Q.

Estimates in the levelling algorithm ([38]).

LEMMA 7.4. *The following equalities hold:*

$$\text{(7.23)}\qquad \begin{aligned} \max_y |f_n(x,y)| &= \max_y |f_{n-1}(x,y)| - |g_n(x)|, \qquad n = 2k+1,\\ \max_x |f_n(x,y)| &= \max_x |f_{n-1}(x,y)| - |h_n(y)|, \qquad n = 2k. \end{aligned}$$

The proof is contained in (6.15) if we recall the definitions of f_n in terms of f_{n-1} (see (7.1)).

LEMMA 7.5. *Let*

$$\Phi(x, y) = \Psi(x, y) - \alpha(x) - \beta(y). \tag{7.24}$$

If Φ *and* Ψ *are vertically (horizontally) levelled functions, then*

$$\|\alpha\| \leq \|\beta\| \qquad (\|\beta\| \leq \|\alpha\|)\,. \tag{7.25}$$

PROOF. Consider the case of vertically levelled Φ and Ψ. Since Φ is vertically levelled, we have

$$\begin{aligned} \alpha(x) =& \frac{1}{2}\left[\max_y (\Psi(x,y) - \beta(y)) + \min_y (\Psi(x,y) - \beta(y))\right] \\ \leq& \frac{1}{2}\left[\max_y \Psi(x,y) + \|\beta\| + \min_y \Psi(x,y) + \|\beta\|\right] = \frac{1}{2}\cdot 2\|\beta\| = \|\beta\|. \end{aligned} \tag{7.26}$$

We have used here that $\max_y \Psi(x, y) = -\min_y \Psi(x, y)$ due to our assumption that Ψ is vertically levelled. Similarly, we have

$$\alpha(x) \geq \frac{1}{2}\left[\max_y \Psi(x,y) - \|\beta\| + \min_y \Psi(x,y) - \|\beta\|\right] = -\|\beta\|. \tag{7.27}$$

From (7.26) and (7.27) it follows that $|\alpha(x)| \leq \|\beta\|$, hence $\|\alpha\| \leq \|\beta\|$.

Monotone decrease of norms of g_n and h_n.

LEMMA 7.6. *For* $k \geq 1$, *the functions* $g_n(x)$ *and* $h_n(y)$ *in the levelling algorithm satisfy the inequalities*

$$\|h_{2k}\| \geq \|g_{2k+1}\| \geq \|h_{2k+2}\|\,. \tag{7.28}$$

PROOF. Consider the functions $\{f_n\}$ in (4.1). Then,

$$f_{2k+1}(x, y) = f_{2k-1}(x, y) - g_{2k+1}(x) - h_{2k}(y) \tag{7.29}$$

and $f_{2,+1}$, f_{2k-1} are vertically levelled. In view of Lemma 7.4 we obtain from (7.29) the first inequality in (7.28). Considering the equality

$$f_{2k+2}(x, y) = f_{2k}(x, y) - g_{2k+1}(x) - h_{2k+2}(y)$$

in which f_{2k+2} and f_{2k} are horizontally levelled, we obtain the second inequality in (7.28). Thus, we have the monotone decrease of the norms

$$\|h_2\| \geq \|g_3\| \geq \|h_4\| \geq \|g_5\| \geq \cdots\,. \tag{7.30}$$

Using Lemma 7.4 we can derive by a similar argument the inequalities

$$\begin{gathered} \|g_3 + g_5 + \cdots + g_{2k+1}\| \leq \|h_2 + h_4 + \cdots + h_{2k}\|\,, \\ \|h_4 + \cdots + h_{2k}\| \leq \|g_3 + \cdots + g_{2k-1}\|\,, \end{gathered}$$

and even more general inequalities for all natural k and $s \leq k - 1$:

$$\begin{gathered} \|g_{2k+1-2s} + g_{2k+1-2s+2} + \cdots + g_{2k+1}\| \leq \|h_{2k-2s} + \cdots + h_{2k}\|\,; \\ \|h_{2k-2s+2} + \cdots + h_{2k}\| \leq \|g_{2k+1-2s} + \cdots + g_{2k-1}\|\,. \end{gathered}$$

LEMMA 7.7. *Let $n \geq 1$, and let x_{2n} be a point in X. If a point $y_{2n-1} \in Y$ is such that*

$$\max_y |f_{2n}(x_{2n}, y)| = |f_{2n}(x_{2n}, y_{2n-1})|, \tag{7.31}$$

then

$$|h_{2n}(y_{2n-1})| \geq 2|g_{2n+1}(x_{2n})| - \|h_{2n}\| \tag{7.32}$$

and

$$\max_x |f_{2n-1}(x, y_{2n-1})| \geq \max_y |f_{2n}(x_{2n}, y)| + |h_{2n}(y_{2n-1})|. \tag{7.33}$$

Let $n \geq 2$ and let y_{2n-1} be a point in Y. If a point $x_{2n-2} \in X$ is such that

$$\max_x |f_{2n-1}(x, y_{2n-1})| = |f_{2n-1}(x_{2n-2}, y_{2n-1})|, \tag{7.34}$$

then

$$|g_{2n-1}(x_{2n-2})| \geq 2|h_{2n}(y_{2n-2})| - \|g_{2n-2}\| \tag{7.35}$$

and

$$\max_y |f_{2n-2}(x_{2n-2}, y)| \geq \max_x |f_{2n-1}(x, y_{2n-1})| + |g_{2n-1}(x_{2n-2})|. \tag{7.36}$$

PROOF. Introduce the notation

$$\begin{gathered} R_{2n} = \max_y |f_{2n}(x_{2n}, y)|, \qquad R_{2n-1} = \max_x |f_{2n-1}(x, y_{2n-1})|, \\ R_{2n-2} = \max_y |f_{2n-2}(x_{2n-2}, y)|, \qquad q_{2n} = |g_{2n+1}(x_{2n})|, \\ q_{2n-1} = |h_{2n}(y_{2n-1})|, \qquad q_{2n-2} = |g_{2n-1}(x_{2n-2})|. \end{gathered} \tag{7.37}$$

Since both statements of the lemma are proved in a similar fashion, we shall concentrate on the first one. First of all, we show that

$$R_{2n} \geq \max_y |f_{2n-1}(x_{2n}, y)| + 2q_{2n} - \|h_{2n}\|. \tag{7.38}$$

According to (7.1) we have

$$q_{2n} = |g_{2n+1}(x_{2n})| = \frac{1}{2}\left|\max_y f_{2n}(x_{2n}, y) + \min_y f_{2n}(x_{2n}, y)\right|. \tag{7.39}$$

Assume that $g_{2n+1}(x_{2n}) > 0$. Then we can drop the absolute value sign in (7.39). We have, furthermore,

$$\begin{aligned} \min_y f_{2n}(x_{2n}, y) &= \min_y [f_{2n-1}(x_{2n}, y) - h_{2n}(y)] \\ &\leq \min_y f_{2n-1}(x_{2n}, y) + \|h_{2n}\| = -\max_y f_{2n-1}(x_{2n-1}, y) + \|h_{2n}\|. \end{aligned} \tag{7.40}$$

The last equality in (7.40) appeared since f_{2n-1} is vertically levelled. From (7.39) and (7.40) we obtain

$$q_{2n} \leq \frac{1}{2}\left[\max_y f_{2n}(x_{2n}, y) - \max_y f_{2n-1}(x_{2n}, y)\right] + \|h_{2n}\|. \tag{7.41}$$

Since again $\max_y |f_{2n-1}(x_{2n},y)| = \max_y f_{2n-1}(x_{2n},y)$ because f_{2n-1} is vertically levelled, formula (7.41) implies (7.38). Similar considerations yield (7.38) for the case when $g_{2n+1}(x_{2n}) < 0$. In view of the choice of y_{2n-1}, we obtain from (7.38)

$$\begin{aligned} R_{2n} - 2q_{2n} + \|h_{2n}t\| &\geq \max_y |f_{2n-1}(x_{2n},y)| \\ &\geq |f_{2n-1}(x_{2n},y_{2n-1})| = |f_{2n}(x_{2n},y_{2n-1}) + h_{2n}(y_{2n-1})| \\ &\geq |f_{2n}(x_{2n},y_{2n-1})| - |h_{2n}(y_{2n-1})| = R_{2n} - q_{2n-1}. \end{aligned}$$

This inequality coincides with (7.38). Furthermore, according to (7.23) we have

$$\begin{aligned} \max_x |f_{2n-1}(x,y_{2n-1})| - |h_{2n}(y_{2n-1})| &= \max_x |f_{2n}(x,y_{2n-1})| \\ &\geq |f_{2n}(x_{2n},y_{2n-1})| = \max_y |f(x_{2n},y)| = R_{2n}, \end{aligned}$$

i.e., we obtain (7.33).

We change the notation to a more symmetric one, denoting $h_{2n}(y)$ by $g_{2n}(y)$, $n = 1,2,\ldots$ (so g_{2n+1} is a function of x, whereas g_{2n} is a function of y).

The main estimate of Diliberto and Straus.

THEOREM 7.8. *For all indices $N > 1$ and $m \geq 0$ the following inequality holds:*

$$\|f\| \geq m\|g_N\| - 2(2^m - 1)(\|g_N\| - \|g_{N+m}\|). \tag{7.42}$$

PROOF. We give a proof for the case when N and m are even. Set $m = 2k$ and $N = 2n - 2k$. The inequality (7.42) becomes

$$\|f\| \geq 2k\|g_{2n-2k}\| - 2(2^{2k} - 1)(\|g_{2n-2k}\| - \|g_{2n}\|). \tag{7.43}$$

We use notation similar to (7.37), keeping in mind that now $g_{2k} = h_{2k}$. Choose a point y_{2n-1} so that $q_{2n-1} = \|g_{2n}\|$. According to Lemma 7.7 there exists a point x_{2n-2} for which (7.35) and (7.36) hold. According to Lemma 7.7, for x_{2n-2} there exists y_{2n-3} such that inequalities (7.32) and (7.33) hold with obvious adjustments of indices. Proceeding in a similar fashion, we obtain points $y_{2n-1}, x_{2n-2}, y_{2n-3}, x_{2n-4}, \ldots, y_{2n-2k-1}$ and numbers $q_{2n-1}, q_{2n-2}, \ldots, q_{2n-2k-1}$; $R_{2n-1}, \ldots, R_{2n-2m-1}$. The inequalities similar to (7.32), (7.33), (7.35), and (7.36) hold for these numbers with obvious adjustments of indices. Let us prove that

$$q_{2n-i} \geq \|g_{2n-2k}\| - 2^{i-1}(\|g_{2n-2k}\| - \|g_{2n}\|). \tag{7.44}$$

We use induction. For $i = 1$ we have

$$q_{2n-1} = \|g_{2n}\| = \|g_{2n-2k}\| - 2^0(\|g_{2n-2k}\| - \|g_{2n}\|).$$

Then (7.44) holds for $i = 1$. Let (7.44) hold for some $i < 2k + 1$. Then, by (7.32) or (7.35) (depending on i being even or odd), we obtain

$$\begin{aligned} q_{2n-i-1} \geq 2q_{2n-i} - \|g_{2n-i}\| &\geq 2\|g_{2n-2k}\| - 2^i(\|g_{2n-2k}\| - \|g_{2n}\|) - \|g_{2n-i}\| \\ &\geq \|g_{2n-2k}\| - 2^i(\|g_{2n-2k}\| - \|g_{2n}\|). \end{aligned} \tag{7.45}$$

We took into account that $\|g_{2n-2k}\| \geq \|g_{2n-i}\|$, according to Lemma 7.6. The inequality (7.45) is obtained from (7.44) by replacing i with $i + 1$, and therefore (7.44) holds for $i = 1, \ldots, k + 1$.

By (7.33) and (7.35) (again making obvious adjustments of indices) we also have:

$$R_{2n-i-1} \geq R_{2n-i} + q_{2n-i-1}, \qquad i = 1, \dots, 2k. \tag{7.46}$$

Using (7.46) and (7.44) we find

$$\begin{aligned} R_{2n-2k-1} \geq & R_{2n-2k} + q_{2n-2k-1} \geq R_{2n-2k+1} + q_{2n-2k-1} + q_{2n-2k} \\ \geq & \cdots \geq R_{2n+1} + \sum_{i=2}^{2k+1} q_{2n-i} \geq \sum_{i=2}^{2k+1} q_{2n-i} \\ \geq & 2k \left\| g_{2n-2k} \right\| - \sum_{i=2}^{2k+1} 2^{i-1} \left(\left\| g_{2n-2k} \right\| - \left\| g_{2n} \right\| \right). \end{aligned} \tag{7.47}$$

Inequality (7.43) follows from (7.47) in view of the obvious inequality $\|f\| \geq \|f_{2n-2k-1}\| \geq R_{2n-2k-1}$.

Tending to zero of $\|g_N\|$.

COROLLARY 7.9. $\lim_{N \to \infty} \|g_N\| = 0$.

PROOF. Take an arbitrary $\varepsilon > 0$ and choose a natural number m so that $\dfrac{\|f\|+1}{m} < \varepsilon$. Since $\|g_N\|$ is monotone decreasing, there exists $\lim \|g_N\|$, so there exists a number N such that $2\left(2^m - 1\right)\left(\|g_N\| - \|g_{N+m}\|\right) \leq 1$. Then, from the inequality (7.42) it follows that

$$\|g_N\| \leq \frac{\|f\|+1}{m} < \varepsilon. \tag{7.48}$$

Convergence of the levelling algorithm. Lemmas.

LEMMA 7.10. *Let $f(x,y) \in C(Q)$, while $G_n(x)$ and $H_n(y)$ are defined by the formulas* (7.3). *Then*

$$G_n(x) = M_y\left[f - H_{n-1}\right](x), \qquad (n \quad \textit{is odd}), \tag{7.49}$$

$$H_n(y) = M_x\left[f - G_{n-1}\right](y), \qquad (n \quad \textit{is even}); \tag{7.50}$$

$$G_2 = G_1,\ G_4 = G_3, \dots; \qquad H_1 = 0,\ H_3 = H_2,\ H_5 = H_4, \dots .$$

In particular, G_n gives the best approximation to $f(x,y) - H_{n-1}(y)$ among all functions $\varphi(x)$ $(n = 2m+1)$, while $H_n(y)$ $(n = 2m)$ gives the best approximation to $f(x,y) - G_{n-1}(x)$ among all functions $\psi(y)$.

PROOF. Let n be an odd number. Then the function $f_n(x,y) = f(x,y) - H_{n-1}(y) - G_n(x)$ is vertically levelled. Yet, in order to level the function $f(x,y) - H_{n-1}(y)$ vertically, the function $G_n(x)$ must be defined as in the formula (7.49). However, in this case, $G_n(x)$ also gives the best approximation to $f - H_{n-1}$ among all functions $\varphi(x)$. The case when n is even is treated similarly. The identities (7.50) are obvious in view of (7.3).

LEMMA 7.11. *Let $G_n(x)$ and $H_n(y)$ be defined by formulas* (7.3). *Let $\varepsilon>0$, $x_1\in X$, and $x_2\in X$ be fixed. If*

$$|f(x_1,y)-f(x_2,y)|<\varepsilon \quad \text{for all } y\in Y, \tag{7.51}$$

then

$$|G_n(x_n)-G_n(x_2)|<\varepsilon. \tag{7.52}$$

Similarly, if $y_1\in Y$ and $y_2\in Y$ are fixed and

$$|f(x,y_1)-f(x,y_2)|<\varepsilon \quad \text{for all } x\in X, \tag{7.51'}$$

then

$$|H_n(y_1)-H_n(y_2)|<\varepsilon. \tag{7.52'}$$

PROOF. We prove the statement regarding $G_n(x)$ (in that case, n is an odd number). In view of (7.51) we have

$$|f(x_1,y)-H_{n-1}(y)-(f(x_2,y)-H_{n-1}(y))|<\varepsilon. \tag{7.53}$$

Consider two functions of y: $v_1=f(x_1,y)-H_{n-1}(y)$ and $v_2=f(x_2,y)-H_{n-1}(y)$. The inequality (7.53) gives

$$\|v_1-v_2\|<\varepsilon.$$

Now apply the estimate (6.14) from Lemma 6.13 and formula (7.49):

$$|G_n(x_1)-G_n(x_2)|=|Mv_1-Mv_2|\le\|v_1-v_2\|<\varepsilon. \tag{7.54}$$

LEMMA 7.12 [**38**]. *There exists a subsequence of functions*

$$f_{n_k}(x,y)=f(x,y)-G_{n_k}(x)-H_{n_k}(y)$$

converging uniformly on Q to the levelled function

$$f^\star(x,y)=f(x,y)-\varphi^\star(x)-\psi^\star(y), \qquad \varphi^\star\in C(X),\quad \psi^\star\in C(Y). \tag{7.55}$$

PROOF. In view of Lemma 7.1 the sequences of functions $\{G_n(x)\}$ and $\{H_n(y)\}$ are equicontinuous. Therefore, the sequence of functions of two variables

$$\{G_n(x)+H_n(y)\}$$

is equicontinuous (in two variables) on $Q=X\times Y$. Since

$$\|G_n+H_n\|\le\|f\|+\|f_n\|\le 2\|f\|,$$

the sequence $\{G_n(x)+H_n(y)\}$ is uniformly bounded. According to Arzelà's theorem (for functions of two variables) there exists a subsequence

$$\{G_{n_k}(x)+H_{n_k}(y)\}$$

uniformly converging on Q to a function $F(x,y)$. In view of Corollary 2.7 $D(Q)$ is closed in $C(Q)$, and therefore $f(x,y)\in D$, i.e., $F(x,y)=\varphi^\star(x)+\psi^\star(y)$.

It remains to show that $f^\star(x,y)=f(x,y)-\varphi^\star(x)-\psi^\star(y)$ is a levelled function. This is obvious when among the sequence of indices $\{n_k\}$ there are infinite sequences of both even and odd indices: indeed, $f^\star=\lim f_{n_k}$, where f_{n_k} is vertically levelled when n_k is odd and horizontally levelled when n_k is even. If $\{n_k\}$ consists, say, exclusively of even indices, we have

$$\lim f_{n_{k+1}} = \lim f_{n_k}$$

because $f_{n_{k+1}} = f_{n_k} - g_{n_{k+1}}$ and $\|g_n\| \to 0$ in view of Corollary 4.8. However, the f_{n_k} are horizontally levelled and the $f_{n_{k+1}}$ are vertically levelled, and we once again arrive at the desired result.

One more monotonicity property for norms of the best approximations.

LEMMA 7.13 [**10**]. *Let* $f_0(x,y) = f(x,y) - \varphi_0(x) - \psi_0(y)$ *be a levelled function (e.g.,* $f_0 = f^\star$ *in Lemma* 7.12*). The following inequalities hold for sequences* $\{G_n(x)\}$ *and* $\{H_n(y)\}$ *defined by* (7.3):

$$\|G_1 - \varphi_0\| \geq \|H_2 - \psi_0\| \geq \|G_3 - \varphi_0\| \geq \|H_4 - \psi_0\| \geq \cdots. \tag{7.56}$$

(*According to* (7.50), $G_1 = G_2$, $G_3 = G_4, \ldots$; $H_2 = H_3, \ldots$.)

PROOF. Let $n \geq 2$ be an even number. Then $f_n(x,y) = f(x,y) - G_n(x) - H_n(y) = f(x,y) - G_{n-1}(x) - H_n(y)$ is a horizontally levelled function. Furthermore,

$$f_n(x,y) = f_0(x,y) + \varphi_0(x) + \psi_0(y) - G_{n-1}(x) - H_n(y), \tag{7.57}$$

where $f_0(x,y)$ is also a horizontally levelled function. According to Lemma 7.5, we obtain the inequality

$$\|G_{n-1} - \varphi_0\| \geq \|H_n - \psi_0\|.$$

Similarly, in case n is odd we show that

$$\|G_n - \varphi_0\| \leq \|H_{n-1} - \psi_0\|.$$

COROLLARY 7.14. *Each one of the sequences* $\{G_n(x)\}$ *and* $\{H_n(y)\}$ *is uniformly bounded.*

Now we can essentially strengthen Lemma 7.12.

The final theorem (Aumann) on the convergence of the algorithm.

THEOREM 7.15 [**10**]. *Let* $Q = X \times Y$. *The sequences* $\{G_n(x)\}$ *and* $\{H_n(y)\}$ *defined in the levelling algorithm by* (7.3) *converge in* $C(Q)$ *to the functions* $\varphi^\star(x)$ *and* $\psi^\star(y)$*, respectively. Moreover, the function* $\varphi^\star(x) + \psi^\star(y)$ *gives the best approximation to* $f(x,y)$ *among all functions* $\varphi(x) + \psi(y)$. *The function* $f^\star(x,y) = f(x,y) - \varphi^\star(x) - \psi^\star(y)$ *is levelled. The functions* $\varphi^\star(x)$ *and* $\psi^\star(y)$ *also satisfy the following uniform continuity properties: given* $\varepsilon > 0$ *and assuming that* (7.51) *holds, we have*

$$|\varphi^\star(x_1) - \varphi^\star(x_2)| < \varepsilon; \tag{7.58}$$

whereas assuming that (7.51′) *holds, we have*

$$|\psi^\star(y_1) - \psi^\star(y_2)| < \varepsilon. \tag{7.58′}$$

PROOF. By Corollary 7.14 and Lemma 7.11 we can, according to Arzelà's theorem, extract uniformly converging subsequences $\{G_{n_k}(x)\}$ and $\{H_{n_k}(y)\}$ from the sequences $\{G_n(x)\}$ and $\{H_n(y)\}$. Let $\varphi^\star(x)$ and $\psi^\star(y)$ be the limits of those subsequences. According to Lemma 7.12, the function $f^\star(x,y)=f(x,y)-\varphi^\star(x)-\psi^\star(y)$ is levelled. In view of the inequalities (7.56) in Lemma 7.13, where we replace φ_0 by $\varphi^\star$ and ψ_0 by $\psi^\star$, we obtain

$$\varphi^\star(x)=\lim_{n\to\infty}G_n(x),\qquad \psi^\star(y)=\lim_{n\to\infty}H_n(y),\tag{7.59}$$

i.e., instead of limits of subsequences, we have limits of the sequences themselves. According to Corollary 7.2 we obtain

$$E(f)=E\left(f^\star\right)=\|f^\star\|\,.$$

Therefore, $\varphi^\star(x)+\psi^\star(y)$ provides the best approximation to $f(x,y)$ in the subspace $D(Q)$. Inequalities (7.58) and (7.58′) follow from Lemma 7.11 and (7.59).

Properties (7.58) and (7.58′) of our best approximant $\varphi^\star(x)+\psi^\star(y)$ can also be expressed as follows: the modulus of continuity of $\varphi^\star(x)$ is not worse than the partial modulus of continuity of $f(x,y)$ with respect to the variable x, and the modulus of continuity of $\psi^\star(y)$ is not worse than the partial modulus of continuity of $f(x,y)$ with respect to y.

Discontinuous functions. Now let $Q\subset X\times Y$, where X and Y are arbitrary sets. For a function $f\in B(Q)$, consider the problem of best approximation:

$$\mathcal{E}(f)=\inf_{\substack{\varphi\in B(X)\\ \psi\in B(Y)}}\|f-\varphi-\psi\|_{B(Q)}.\tag{7.60}$$

As an attempt to solve this problem, we may try to set up the same levelling algorithm described by the formulas (7.1). Max and min in these formulas now have to be replaced by sup and inf.

THEOREM 7.16. *Let Q contain a bar. Then the statement of Theorem* 7.1 *and formula* (7.8) *with $E(f)$ replaced by $\mathcal{E}(f)$ hold for $f\in B(Q)$. Any function $F(x,y)$ that is a limit point of a sequence $\{G_n(x)+H_n(y)\}$ in the weak $(\star)$ topology of the space $B(Q)$ belongs to $BD(Q)$ and gives the best approximation in the problem* (7.60).

PROOF. For definiteness, assume that Q contains a horizontal bar passing through the point (x_0,y_0). As in the proof of Theorem 7.1, construct a sequence of points $(x_1,y_1),(x_1,y_2),(x_2,y_2),\dots,(x_{m+1},y_{m+1})$. Form a closed lightning bolt ℓ out of the points $(x_1,y_0),(x_1,y_2),\dots,(x_{m+1},y_{m+1}),(x_{m+1},y_0)$. The rest is as in Theorem 7.1. Now let $F(x,y)$ be a limit point of the sequence $\{G_n(x)+H_n(y)\}$ in the weak $(\star)$ topology of $B(Q)$. (Since $f_n=f-G_n-H_n$, we have $\|G_n+H_n\|\le 2\|f\|$; hence the sequence $\{G_n+H_n\}$ has a limit point.) For any closed lightning bolt L we then have $r_L(F)=0$, because $r_L\left(G_n+H_n\right)=0$. If $\psi(x,y)=F(x,y)-F\left(x,y_0\right)$, then $r_L(\psi)=r_L(F)-r_L\left(F\left(x,y_0\right)\right)=0$. Taking two points and constructing a closed lightning bolt L: (x_1,y_0), (x_1,y), (x_2,y), (x_2,y_0), we find that $\psi\left(x_1,y\right)=\psi\left(x_2,y\right)$, and so ψ only depends on y. Thus, $F(x,y)=\varphi(x)+\psi(y)$ $(\varphi(x)=F\left(x,y_0\right))$. Since $f-F$ is a weak $(\star)$ limit point for $f-G_n-H_n$, then $\|f-F\|\le\overline{\lim}\,\|f-G_n-H_n\|=\mathcal{E}(f)$ and hence F provides the best approximation to f in $BD(Q)$.

Estimate of the rate of decrease of $\|g_n\|$. Again, we keep the notation $g_{2n+1}(x)$, $g_{2n}(y)$ for the functions in (7.1).

THEOREM 7.17. *For any $\varepsilon > 0$ there exists n_0 such that for $n > n_0$*

$$\|g_n\| < \frac{(1+\varepsilon)\|f\|}{\log_2 n}. \tag{7.61}$$

PROOF. Assume that (7.61) is false. Then there exist $\varepsilon > 0$ and an infinite set of indices J such that

$$\|g_n\| \geq \frac{(1+\varepsilon)\|f\|}{\log_2 n}, \qquad n \in J. \tag{7.62}$$

We take ε_1, $0 < \varepsilon_1 < \varepsilon$, and show that there exist sequences $k_j \to \infty$ and m_j such that

$$\|g_{k_j}\| \geq [(1-\varepsilon)\|f\| - O(1)]/\log_2 k_j \tag{7.63}$$

and

$$\|g_{k_j}\| - \|g_{k_j+m_j}\| = O(k_j^{1/(1+\varepsilon_1)}), \tag{7.64}$$

where

$$\left| m_j - \frac{\log_2 k_j}{1+\varepsilon_1} \right| \leq 2.$$

If such sequences are constructed, then taking $N = k_j$, $m = m_j$ in the inequality (7.42) we obtain

$$\begin{aligned} \|f\| >&[\frac{\log_2 k_j}{1+\varepsilon_1} - 2][(1+\varepsilon)\|f\| - O(1)]/\log_2 k_j \\ &- 2(2^2 k_j^{1/(1+\varepsilon_1)} - 1)O(k_j^{-1/(1+\varepsilon_1)}). \end{aligned} \tag{7.65}$$

Clearly, when $k_j \to \infty$ the right-hand side in (7.65) tends to $\dfrac{(1+\varepsilon)\|f\|}{1+\varepsilon_1} > \|f\|$. This contradiction proves the theorem.

Thus, everything is reduced to constructing $\{k_j\}$ and $\{m_j\}$. Choose an arbitrary index N, and choose $n > N$ so that $n \in J$. Insert between N and n a finite number of indices $q_1, \dots, q_M$ as follows. Set $q_1 = n$. If $q_1 > q_2 > \cdots > q_i$ are constructed, choose q'_{i+1} such that

$$q'_{i+1} + \frac{\log_2 q'_{i+1}}{1+\varepsilon_1} = q_i. \tag{7.66}$$

After that, set $q_{i+1} = [q'_{i+1}]+1$ ($[\,]$ denotes the integral part). Terminate the process at a point q_M such that $q_{M+1} < N \leq q_M$. Changing the notation, renumber all the natural numbers q_j according to their growth: $N \leq p_1 < \cdots < p_M = n$. In view of (7.66) it is clear that

$$0 \leq p_i + \frac{\log_2 p_i}{1+\varepsilon_1} - p_{i+1} \leq 2. \tag{7.67}$$

From (7.67) we obtain

$$p_{i+1} - p_i < \log_2 p_i \leq \log_2 n. \tag{7.68}$$

Hence

$$M > (n-N)\log_2 n \quad \text{and} \quad p_i < N + i\log_2 n. \tag{7.69}$$

Choose ε_2, $0<\varepsilon_2<\varepsilon_1$. Let

$$p_{i+1}\in J, \qquad \|g_{p_i}\| - \|g_{p_{i+1}}\| \ge p_i^{-1/(1+\varepsilon_2)}. \tag{7.70}$$

Then from (7.62) and (7.70) we obtain for $N > N_0(\varepsilon_2)$

$$\|g_{p_i}\| \ge \frac{(1+\varepsilon)\|f\|}{\log_2 p_{i+1}} + p^{-1/(1+\varepsilon_2)} > \frac{(1+\varepsilon)\|f\|}{\log_2 p_i}. \tag{7.71}$$

Indeed, for $N > N_0(\varepsilon_2)$ using (7.68) we obtain

$$\begin{aligned}(1+\varepsilon)\|f\|\left(\frac{1}{\log_2 p_i} - \frac{1}{\log_2 p_{i+1}}\right) &< (1+\varepsilon)\|f\|\log_2\frac{p_{i+1}}{p_i}/\log_2^2 p_i\\ &<(1+\varepsilon)\|f\|\log_2 e\cdot\frac{p_{i+1}-p_i}{p_i}/\log_2^2 p_i\\ &<(1+\varepsilon)\|f\|\log_2 e\cdot\log_2 p_i/p_i\log_2^2 p_i < p_i^{-1/(1+\varepsilon_2)}.\end{aligned}$$

So, if $p_{i+1}\in J$ and (7.70) holds, then $p_i\in J$. Since $p_M = n\in J$, we have the following alternative: either

(a) all numbers $p_1,\dots,p_M$ belong to J, or
(b) there exists p_{i+1} such that

$$\|g_{p_i}\| - \|g_{p_{i+1}}\| \le p_i^{-1/(1+\varepsilon_2)}, \qquad p_{i+1}\in J. \tag{7.72}$$

Let us show that for sufficiently large n (a) is impossible. Otherwise, adding all the inequalities (7.70) we would have obtained using (7.69) ($\beta=-1/(1+\varepsilon_2)$):
(7.73)

$$\begin{aligned}\|g_{p_1}\| &\ge \sum_{i=1}^{M-1} p_i^{\beta} > \sum_{i=1}^{M-1}(N+i\log_2 n)^{\beta} > \int_1^M (N+x\log_2 n)^{\beta}\,dx\\ &=(\beta+1)^{-1}(\log_2 n)^{-1}\left\{[N+M\log_2 n]^{\beta+1} - [N+\log_2 n]^{\beta+1}\right\}\\ &\ge(\beta+1)^{-1}(\log_2 n)^{-1}\left\{1-\left[\frac{N+\log_2 n}{N+M\log_2 n}\right]^{\beta+1}\right\}[n+(n-N)]^{\beta+1}.\end{aligned}$$

As $n\to\infty$, the expression in braces in (7.73) tends to 1, whereas the entire right-hand side of (7.73) tends to ∞. Thus, for sufficiently large n the inequality (7.73) yields a contradiction. Therefore, for all sufficiently large $n>N$ the possibility (b) holds. Take a sequence of indices $\{N^j\}\uparrow+\infty$. In view of our construction, there exists a sequence $\{p_i^j\}\uparrow+\infty$ such that

$$p_i^j\ge N^j, \qquad p_{i+1}^j\in J, \quad \left\|g_{p_i^j}\right\| - \left\|g_{p_{i+1}^j}\right\| < \left(p_i^j\right)^{-1/(1+\varepsilon_2)}. \tag{7.74}$$

Set $k_j = p_i^j$, $m_j = p_{i+1}^j - p_i^j$. In view of (7.74) and the fact that $p_{i+1}^j/p_i^j\to 1$ (due to (7.68)), we obtain the inequalities (7.63). From (7.74) we also get the first relation in (7.64). The second inequality in (7.64) follows from (7.67). The theorem is proved.

Bibliographical notes. The levelling algorithm for $f \in C(X \times Y)$ and $f \in B(X \times Y)$ was studied in the paper [**38**] by Diliberto and Straus. Strictly speaking, they considered $X = [a, b] \subset \mathbb{R}^1$ and $Y = [c, d] \subset \mathbb{R}^1$, but this is not important. As noted in [**38**], the question of studying the problem of the best approximation of a function of two variables by functions $\varphi(x) + \psi(y)$ was (in a slightly different form) posed by Rand Corporation. In that paper, the authors proved Theorem 7.1 and the formula (7.9). Convergence of a sequence of functions obtained along the steps of the algorithm to functions that solve the best approximation problem and simultaneously levelling the initial function was established somewhat later by Aumann [**10**]. Diliberto and Straus only showed convergence of subsequences. Aumann [**9–11**] had arrived at the study of this approximation problem independently of [**38**]. Also, in [**38**] the authors showed that the best approximation $\varphi^\star(x) + \psi^\star(y)$ obtained by the levelling process for a continuous function $f(x, y)$ has continuity properties that are not worse than those of the function itself.

We have already pointed out in §3,6 that the existence of the best approximation $\varphi^\star(x) + \psi^\star(y)$ and continuity properties of $\varphi^\star(x) + \psi^\star(y)$ for $f \in C(X \times Y)$ had also been established by Kolmogorov (cf. the exposition in [**113**]). Although [**113**] contains no systematic study of the levelling algorithm, this very natural idea can be clearly seen in the proofs in [**113**].

In [**38**] there are given the estimates (7.42) and

$$\|g_n\| \leq (2 + \varepsilon)\|f\| / \log_2 n, \quad \text{for all } \varepsilon > 0,\ n > n_0(\varepsilon). \tag{7.75}$$

Both estimates are given without proof. (The authors point out that the proofs were omitted in accordance with the referee's suggestion, in view of the possibility that those estimates were not sharp and yet required tedious calculations.) The proofs of (7.42) and (7.61) given here are due to Eiderman [**43**]; the estimate (7.61) is somewhat better than (7.75). In [**38**] it was conjectured that

$$\|g_n\| \leq C2^{-n}, \tag{7.76}$$

but this is false (at least for bounded functions). In [**91**] an example is given for which the estimate (7.76) does not hold. In the preprint [**44**] Eiderman proved the following estimate for bounded functions on an arbitrary set $Q \subset \mathbb{R}^2$:

$$\|g_{2n+1}\| \leq \|f\| \binom{2n}{n} 2^{-2n} \sim \|f\|(\pi n)^{-\frac{1}{2}} \tag{7.77}$$

(the same asymptotic estimate holds for $\|g_{2n+2}\|$) and constructed a bounded function for which equality holds in (7.77) for all $n = 1, 2, \dots$. We chose not to present these cumbersome results here.

In addition to the papers quoted in this section, let us mention the papers [**64**], [**56**], [**3**], [**132**], and [**23**]. The monograph [**94,** Chapter 5] studies the levelling algorithm but without estimates of the error terms (formulas (7.42), (7.61), and (7.77)). Chapters 3,4,6,7 of [**94**] contain results (both positive and negative) on generalizations of the levelling algorithm in other metrics, together with an extensive bibliography.

In [**38**], Diliberto and Straus described (without detailed proofs; "by similarity") a "natural" extension of the levelling algorithm to several variables. A similar extension is suggested in Golomb's paper [**64**] in a more abstract situation. However, Aumann [**13**] showed that in three variables the levelling algorithm need not lead to the best approximation. Unfortunately, this important negative result is not contained in [**94**]. Unaware of Aumann's work, Medvedev [**104**] also showed that when the number of variables $n > 2$, the Diliberto–Straus algorithm does not always lead to the desired goal. (I learned about Aumann's work from Professor Pinkus at the Haifa conference in 1994, and use this opportunity to express my sincere gratitude to him; cf. §3 in Chapter 3 below.)

CHAPTER 3

Problems of Approximation by Linear Superpositions

§1. Properties of the subspace of linear superpositions and its annihilator

Types of linear superpositions. Let $X, X_1, \ldots, X_N$ be compact sets, and $\Phi_i : X \to X_i$ continuous mappings (cf. (1.1), §1, Chapter 2). Also, assume there are given functions $h^i(x) \in C(X)$. We consider in $C(X)$ a subspace $D(X) = D$ of functions

$$h^1(x)\left[g_1 \circ \Phi_1(x)\right] + \cdots + h^N(x)\left[g_N \circ \Phi_N(x)\right], \tag{1.1}$$

where $g_i \in C(X_i)$, $i = 1, \ldots, N$, are arbitrary functions. The functions (1.1) will be called linear superpositions. In $C(X)$, consider the problem of approximation by functions in the subspace D. If $X, X_1, \ldots, X_N$ are arbitrary sets and $\Phi_i : X \to X_i$, $i = 1, \ldots, N$, are mappings, then instead of continuous functions we may consider bounded ones, and then in (1.1) $h^i(x) \in B(X)$, $g_i(x) \in B(X_i)$ and we denote the subspace of functions (1.1) by BD. In the subspace $B(X)$ consider the problem of approximation by functions in BD. For the moment we shall discuss the situation with continuous functions, indicating the needed alterations for the case of bounded functions at the end of the section.

Consider the product of compact sets X_i:

$$Y = X_1 \times \cdots \times X_N. \tag{1.2}$$

If the mapping

$$\Psi : X \to Y, \qquad \Psi(x) = (\Phi_1(x), \ldots, \Phi_N(x)) \tag{1.3}$$

is injective (or, equivalently, the system $(\Phi_1(x), \ldots, \Phi_N(x))$ separates points in X), the subspace D consists of functions

$$h^1(x)\left[g_1 \circ \pi_1(x)\right] + \cdots + h^N(x)\left[g_N \circ \pi_N(x)\right] = \sum_{i=1}^{N} h^i(x) g_i(x_i), \tag{1.4}$$

where $x = (x_1, \ldots, x_N)$ is a point in Y, $\pi_i(x) = x_i$ is the natural projection of Y onto X_i, $g_i \in C(X_i)$ are arbitrary, $h^i(x) \in C(Y)$. We shall sometimes call the sets X_i a basis.

The approximation problem then reduces to the following: in the space $C(Q)$, where $Q \subset Y$ $(Q = \Psi(X))$, study approximation by functions of the subspace D

(i.e., by functions (1.4)). Let $j = 1, \dots, m$ be some indices, each corresponding to a set (j) of indices i:

$$(j) = (i_1, i_2, \dots, i_{k_j}), \qquad 1 \le i_1 < \dots < i_{k_j} \le N. \tag{1.5}$$

Form partial (with respect to (1.2)) products

$$Y_j = Y_{(j)} = X_{i_1} \times \dots \times X_{i_{k_j}}. \tag{1.6}$$

Denote by $y_j = (x_{i_1}, \dots, x_{i_{k_j}})$ points in Y_j. If $h^k(x) \in C(Y)$ are given functions, then the subspace D of functions

$$\sum_{j=1}^{m} h^j(x) g_j(x_{i_1}, \dots, x_{i_{k_j}}), \qquad g_j \in C(Y_j), \tag{1.7}$$

is of a more general nature than (1.4). The space (1.4) is obtained from (1.7) by letting $m = N$, $(j) = j$, $Y_j = X_j$. The functions (1.7) (in particular, those in (1.4)) are also called linear superpositions. The functions $h^j(x)$ will sometimes be called basis functions, and the g_j will be called the coefficients in linear superpositions (1.1), (1.4), or (1.7). The representation (1.7) for a given function $w \in D$ is, in general, not unique.

Proper bases. A system of basis functions $\{h^j(x)\}$, $j = 1, \dots, m$, in (1.7) is called *proper* in $C(Q)$ if there exists a number $c > 0$ such that for any function $w(x)$ of the form (1.7) the coefficients g_j in the representation (1.7) of w on Q can be chosen so that the inequality

$$\sum_{j=1}^{m} \|g_j\| \le c\|w\|_Q \tag{1.8}$$

holds.

PROPOSITION 1.1. *A system of basis functions is proper in $C(Q)$ if and only if the subspace $D(Q)$ is closed in $C(Q)$.*

PROOF. Introduce another norm in $D(Q)$ by setting

$$|w|_Q = \inf \sum_{j=1}^{m} \|g_i\|, \tag{1.9}$$

where the infimum is taken over all choices of $\{g_j\}$ that provide the representation (1.7) for w on Q. It is easy to check that $|w|_Q$ is indeed a norm. In particular, to check that $|w|_Q = 0 \Leftrightarrow w = 0$, one has to use the estimate $\|w\|_Q \le C|w|_Q$, where $C = \max \|h^j\|$. That $D(Q)$ is complete with respect to the norm (1.9) can be proved along the same scheme as, e.g., the proof of completeness of quotient spaces. For completeness we sketch the arguments here.

Let $\{w_n\} \subset D$ be a fundamental sequence with respect to the norm (1.9). We can always extract a subsequence w_{n_k} so that the series

$$\sum_{k=1}^{\infty} |w_{n_{k+1}} - w_{n_k}|_Q < +\infty.$$

Let functions $g_{n_1,j}$, $j = 1, \ldots, m$, be taken from the representation (1.7) for w_{n_1}, while functions $\varphi_{n_k,j}$, $j = 1, \ldots, m$, are taken from that for $w_{n_{k+1}} - w_{n_k}$. We can assume that

$$\sum_{k=1}^{\infty} |\varphi_{n_k,j}| < +\infty, \qquad j = 1, \ldots, m.$$

Then the series

$$g_{n_1,j} + \sum_{k=1}^{\infty} \varphi_{n_k,j} = g_j,$$

$j = 1, \ldots, m$, converge in $C(Y_j)$ to functions $f_j \in C(Y_j)$. If $w \in D(Q)$ is a function representable in the form (1.7) with the coefficients g_j, then $|w - w_{n_k}|_Q \to 0$ and hence $|w - w_n|_Q \to 0$. This proves that $D(Q)$ is a Banach space with respect to the norm (1.9).

Now let $D(Q)$ be closed in $C(Q)$. The identity operator mapping $(D, |\cdot|_Q)$ into $(D, \|\cdot\|_Q)$ is continuous and according to the Banach theorem so is its converse, i.e., the operator from $(D, \|\cdot\|_Q)$ into $(D, |\cdot|_Q)$, which is equivalent to (1.8), and so the system of basis functions is proper in $C(Q)$. Let us show the converse: a system of basis functions is proper. We need to show that $D(Q)$ is closed in $C(Q)$. Let $\{w_n\} \subset D$ and $w_n \to w$ in $C(Q)$. Since the system $\{h^j\}$ is proper, by (1.8) it follows that $\{w_n\}$ is a fundamental sequence with respect to the norm $|\cdot|_Q$. Hence it converges with respect to this norm to a limit $w_0 \in D$; but then $\{w_n\} \to w_0$ in $C(Q)$ as well. Hence $w = w_0$, and the proposition is proved.

Examples of proper and improper systems. In $C(X_1 \times X_2)$ consider the subspace D consisting of functions $g_1(x_1) + g_2(x_2)$. Set $h^1(x) = h^2(x) = 1$ in (1.1). From §2 in Chapter 2 it follows that the subspace D is closed, and therefore the system $\{1, 1\}$ is proper. Now take $N = 1$, $x_1 = [0, 2]$, and $h^1(x) = 0$ for $1 \leq x \leq 2$, $h^1(x) > 0$ when $0 \leq x < 1$. The subspace $D = \{h^1(x) g_1(x)\}$ with the basis function $h^1(x)$ gives an example of a linearly independent but improper system: there exist functions $g_1(x)$ with an arbitrary large norm so that $\|h^1 g_1\| \leq 1$. This example also shows that for linearly independent basis systems the properness requirement does not reduce to the well-known equivalence of all norms on a finite-dimensional space. Yet, if the variables in the basis functions h^j and functions g_j (see below) are separated, linear independence implies properness of the system (Lemma 1.10 below.).

Separated variables. The most common case is when a basis function $h^j(x)$ only depends on those coordinates of a point x that are not used in forming points y_j in Y_j. Denote by $(\tilde{j})$ the complementary set of indices to (j). So, $(\tilde{j})$ consists of all indices i, $1 \leq i \leq N$, that are not used in (j). By $\widetilde{Y}_j = Y_{(\tilde{j})}$ denote the product of all X_i for which $i \in (\tilde{j})$; $\tilde{y}_j$ are points in $\widetilde{Y}_j$. Then

$$Y_j \times \widetilde{Y}_j = Y, \tag{1.10}$$

provided that from now on we agree on the following. In the expansion of the product (1.10) factors X_i in Y_j and $\widetilde{Y}_j$ are written in the same order as they appear in Y (i.e., in order of increase of indices i) independently of whether they belong to X_j in Y_j, or in $\widetilde{Y}_j$. Similarly, when writing $x = (y_j, \tilde{y}_j) = (\tilde{y}_j, y_j)$, the coordinates x_i are written in the natural order without paying attention to whether a coordinate is taken from y_j, or $\tilde{y}_j$. For example, if $y_j = (x_2, x_5, x_6)$, $\tilde{y}_j = (x_1, x_3, x_4, x_7)$, then $(y_j, \tilde{y}_j) = (\tilde{y}_j, y_j) = (x_1, x_2, x_3, x_4, x_5, x_6, x_7)$.

Let us denote by $\pi_{(j)}$ a natural projection of Y onto $Y_j = Y_{(j)}$ and by $\tilde{\pi}_{(j)} = \pi_{(\tilde{j})}$ a natural projection on $\widetilde{Y}_j$. Thus, if $x = (y_j, \widetilde{y}_j)$, then $\pi_{(j)}(x) = y_j$, $\tilde{\pi}_{(j)}(x) = \pi_{(\tilde{j})}(x) = \tilde{y}_j$. Let $H^j = H^j(\widetilde{Y}_j)$, $j = 1, \dots, m$, be a finite-dimensional subspace in $C(\widetilde{Y}_j)$. Consider in $C(Y)$ a subspace D defined by

$$D = \sum_{j=1}^{m} H^j \otimes C(Y_j). \tag{1.11}$$

Here, $\otimes$ denotes the tensor product, while again, when forming from the functions $f_1(\tilde{y}_j) \in H^j$ and $f_2(y_j) \in C(Y_j)$ a function $F(x) = f_1 \otimes f_2 = f_2 \otimes f_1 \in C(Y)$, coordinates of a point x appear in their natural order irrespectively to whether a coordinate comes from y_j, or $\tilde{y}_j$. Therefore, in particular, for us $f_1(\tilde{y}_j) \otimes f_2(y_j) = f_2(y_j) \otimes f_1(\tilde{y}_j)$ and $H^j(\widetilde{Y}_j) \otimes C(Y_j) = C(Y_j) \otimes H^j(\widetilde{Y}_j)$. From now on whenever we form tensor products we shall follow these rules.

The formula (1.11) even covers a more general situation when (j) and $(\tilde{j})$ in it do not necessarily compliment each other, but are simply disjoint sets of indices. To see that, it suffices to assume that functions in H^j in fact depend only on some of the variables forming $Y_{(\tilde{j})}$. Let $h_j^\delta(\tilde{y}_j)$, $\delta = 1, \dots, n_j$, be a basis of the subspace H^j, $j = 1, \dots, m$. Then the subspace (1.11) consists of the functions

$$\sum_{j=1}^{m} \sum_{\delta=1}^{n_j} h_j^\delta(\tilde{y}_j)\, g_j^\delta(y_j), \qquad g_j^\delta(y_j) \in C(Y_j). \tag{1.12}$$

The functions (1.12) are a special case of the functions (1.7).

The functions $\{h_j^\delta\}$, $j = 1, \dots, m$, $\delta = 1, \dots, n_j$, serve as basis functions for the subspace (1.12). In each term of (1.12) the arguments y of a basis function $h_j^\delta(\tilde{y}_j)$ and of a coefficient $g_j^\delta(y_j)$ are separated. We call this case the case with separated variables.

Totally separated variables. In many situations it is convenient to admit the following structure for the basis subspaces H^j. There are finite-dimensional subspaces $H^i \subset C(X_i)$, $i = 1, \dots, N$, and if $(j) = (i_1, \dots, i_{\tilde{k}_j})$, i.e., $\widetilde{Y}_j = X_{i_1} \times \dots \times X_{i_{\tilde{k}_j}}$, then

$$H^j\left(\widetilde{Y}_j\right) = H^{i_1} \otimes H^{i_2} \otimes \dots \otimes H^{i_{\tilde{k}_j}}. \tag{1.13}$$

The case when the $H^j(\widetilde{Y}_j)$ have this structure is called the case of totally separated variables.

The following situation is even more general. Let $Z_1, \dots, Z_s$ be some partial products out of the product (1.2) and moreover, any two of them, Z_k and Z_ℓ, consist of different factors. Let H^k be a finite-dimensional subspace in $C(Z_k)$, $k = 1, \dots, s$. Further, assume that each $\widetilde{Y}_j$ is a product of some of the Z_k, whereas $H(\widetilde{Y}_j)$ is the tensor product of the corresponding H^k:

$$\widetilde{Y}_j = Z_{k_1} \times \dots \times Z_{k_{n_j}}, \qquad H^j(\widetilde{Y}_j) = H^{k_1} \otimes \dots \otimes H^{k_{n_j}}. \tag{1.14}$$

Clearly, the seemingly greater generality of (1.14) compared to (1.13) is an illusion. If, from the very beginning, while forming products (1.2) one uses the blocks Z_k instead of the factors X_i then (1.14) coincides with (1.13). The case (1.14) will also be called the situation with totally separated variables.

The annihilator of the subspace of linear superpositions.

LEMMA 1.2. *Let D consist of functions* (1.1). *In order that a regular Borel measure μ on X belong to $D^\perp$, it is necessary and sufficient that*

$$\nu_i = \Phi_i \circ \left[h^i(x)\mu\right] \equiv 0, \qquad i = 1, \dots, N. \tag{1.15}$$

In essence, this lemma repeats Lemma 4.1 in Chapter 2. A special case of Lemma 1.2 is the following lemma.

LEMMA 1.3. *Let D consist of functions* (1.7) *and $\mu \in C(Q)^\star$. In order that $\mu \in D(Q)^\perp$, it is necessary and sufficient that*

$$\nu_j = \pi_{(j)} \circ \left[h^j(x)\mu\right] \equiv 0, \qquad j = 1, \dots, m. \tag{1.16}$$

Another characterization of $D^\perp$ when the variables are separated. In case of separated variables, a condition for a measure to belong to $D^\perp$ can be expressed in a different form. Denote by $\mu\,|_E$ the restriction of a measure μ to a set E.

LEMMA 1.4 ([**93**]). *Let the subspace D have the structure* (1.11). *In order that $\mu \in [H^j(\widetilde{Y}_j) \otimes C(Y_j)]^\perp$, it is necessary and sufficient that for any Borel set $A \subset Y_j$ the measure*

$$\nu_j^A = \pi_{(\tilde{j})} \circ \left[\mu\,\Big|_{A \times \widetilde{Y}_j}\right] \in H^j(\widetilde{Y}_j)^\perp. \tag{1.17}$$

PROOF. For a function $h\,(\tilde{y}_j) \in C(\widetilde{Y}_j)$ we have

$$\int_{A \times \widetilde{Y}_j} h d\mu = \int_{\widetilde{Y}_j} h d\nu_j^A. \tag{1.18}$$

Now if $h \in H^j(\widetilde{Y}_j)$ and $\mu \in [H^j(\widetilde{Y}_j) \otimes C(Y_j)]^\perp$, then $\pi_{(j)} \circ [h\mu] \equiv 0$ and hence

$$\int_{A \times \widetilde{Y}_j} h d\mu = \pi_{(j)}\,[h\mu]\,(A) = 0. \tag{1.19}$$

Therefore,

$$\int_{\widetilde{Y}_j} h d\nu_j^A = 0, \quad \text{i.e. } \nu_j^A \in \left(H^j\right)^\perp. \tag{1.20}$$

Conversely, if $\nu_j^A \in \left(H^j\right)^\perp$ for all $A \subset Y$, the integrals in (1.18) vanish, and hence $\pi_{(j)}\,[h\mu]\,(A) = 0$ for all A and $\mu \in \left[H^j \otimes C\,(Y_j)\right]^\perp$.

Product-measures in $D^\perp$ when the variables are separated.

LEMMA 1.5. *Let a subspace D be defined by* (1.11) *and* (1.13). *Let*

$$X_{k_1}, X_{k_2}, \dots, X_{k_s} \tag{1.21}$$

be a collection of basis sets X_i, $i = 1, \dots, N$, so that each product $\widetilde{Y}_j$, $j = 1, \dots, m$, includes at least one set (1.20) *as a factor. If measures $\lambda_{k_\ell} \in C(X_{k_\ell})^\star$, $\ell = 1, \dots, s$, satisfy $\lambda_{k_\ell} \in H^{k_\ell}(X_{k_\ell})^\perp$ and λ is an arbitrary measure defined on a product of sets that are not included in* (1.21), *then*

$$\mu = \lambda_{k_1} \otimes \lambda_{k_2} \otimes \dots \otimes \lambda_{k_s} \otimes \lambda \in D^\perp. \tag{1.22}$$

PROOF. Consider the subspace

$$D_j = H^j \otimes C(Y_j), \tag{1.23}$$

where $\widetilde{Y}_j$ and $H^j(\widetilde{Y}_j)$ are defined by (1.13). Let $\{h_{i_1}^\alpha(x_{i_1}), \alpha = 1, \dots, n_{i_1}\}$ be a basis in $H^{i_1}(X_{i_1}), \dots, \{h_{i_{\bar{k}_j}}^\beta, \beta = 1, \dots, n_{i_{\bar{k}_j}}\}$ be a basis in $H^{i_{\bar{k}_j}}(X_{i_{\bar{k}_j}})$. A basis in $H^j(\widetilde{Y}_j)$ generates various products

$$h_j^\delta(\tilde{y}_j) = h_{i_1}^\alpha(x_{i_1}) \otimes \dots \otimes h_{i_{\bar{k}_j}}^\beta(x_{i_{\bar{k}_j}}) \tag{1.24}$$

and D_j consists of all sums

$$\sum_\delta h_j^\delta(\tilde{y}_j) g_\delta(y_j), \qquad g_\delta \in C(Y_j). \tag{1.25}$$

To fix the ideas, assume that the set X_{i_1} is included into (1.21), e.g., $X_{i_1} = X_{k_1}$. Then $\lambda_{k_1}(h_{i_1}^\alpha) = \lambda_{i_1}(h_{i_1}^\alpha) = 0$ by the hypothesis. Moreover, we have

$$\begin{aligned}
&\lambda_{k_1} \otimes \dots \otimes \lambda_{k_s} \otimes \lambda[h_{i_1}^\alpha \otimes \dots \otimes h_{i_{\bar{k}_j}}^\beta] \\
=&\lambda_{k_1}(h_{i_1}^\alpha)[\lambda_{k_2} \otimes \dots \otimes \lambda_{k_s} \otimes \lambda(h_{i_2}^\alpha \otimes \dots \otimes h_{i_{\bar{k}_j}}^\beta \times g_\delta)] = 0.
\end{aligned}$$

Obviously, Lemma 1.5 can be deduced from Lemma 1.4. For example, for any $A \subset X_{k_1}$ we have

$$\pi_{k_1} \circ [\cdot \mu|_{A \times \widetilde{X}_{k_1}}] = \lambda_{k_1}[\lambda_{k_2} \otimes \dots \otimes \lambda_{k_s} \otimes \lambda](A).$$

Thus, we obtained a measure that differs from λ_{k_1} only by a constant factor and hence is orthogonal to $H^{k_1}(X_{k_1})$.

"Domino"-measures. If each measure in the product (1.22) has a finite support, then we shall call such a product a "domino"-measure (the term was suggested in [**93**]). The simplicity of the structure of "domino"-measures makes them attractive, whereas the fact that they form a sufficiently rich set is ensured by the following proposition.

PROPOSITION 1.6. *Let a subspace D be of the form*

$$D = \sum_{i=1}^{N} H^i(X_i) \otimes C(\tilde{X}_i), \tag{1.26}$$

where the $H^i(X_i)$ are finite-dimensional subspaces in $C(theX_i)$ (X_i, $i = 1, \dots, N$, are basis sets). In order that $w \in D$ it is necessary and sufficient that $\mu(w) = 0$ for any "domino"-measure μ:

$$\mu = \lambda_1 \otimes \cdots \otimes \lambda_N, \tag{1.27}$$

where $\lambda_i \in H^i(X_i)^\perp$, $i = 1, \dots, N$.

Necessity is contained in Lemma 1.5. Sufficiency will be proved later on.

Approximation of measures in $D^\perp$ by measures with a finite support. As usual, to avoid extra notation we shall denote by the same letter a linear functional and the measure representing it. Together with the usual characteristic of a measure μ

$$\|\mu\|_D = \sup_{\substack{f \in D \\ \|f\| \leq 1}} |\mu(f)|, \tag{1.28}$$

its action on D can be estimated by the quantity

$$|\mu|_D = \sup_{\substack{f \in D \\ |f| \leq 1}} |\mu(f)|. \tag{1.29}$$

Since $\|f\| \leq C|f|$, we always have $|\mu| \leq \dfrac{1}{C}\|\mu\|$, and in case of a proper basis system the converse inequality also holds with an appropriate constant.

PROPOSITION 1.7. *Let D be a subspace in $C(Y)$ that consists of functions (1.7), $\mu \in D^\perp$, $\|\mu\| = 1$, and μ^+ (μ^-) the positive (negative) variations of μ. There exist nets of measures $\{\mu_\theta^+\}$ and $\{\mu_\theta^-\}$ with the following properties:*

1. *μ_θ^+, μ_θ^- are positive measures with finite supports and, moreover, $\|\mu_\theta^+\| + \|\mu_\theta^-\| = 1$.*
2. *$\{\mu_\theta^+\}$ ($\{\mu_\theta^-\}$) converges in the weak ($\star$) topology of $C(Y)^\star$ to μ^+ (μ^-).*
3. *There exist nets of functions $\{h_\theta^{j+}\} \subset C(Y)$ and $\{h_\theta^{j-}\} \subset C(Y)$, $j = 1, \dots, m$, each of which converges in $C(Y)$ to the function h^j and*

$$\pi_{(j)} \circ [h_\theta^{j+} \mu_\theta^+] = \pi_{(j)} \circ [h_\theta^{j-} \mu_\theta^-]. \tag{1.30}$$

4. *For each $\varepsilon > 0$ there exists $\theta_0(\varepsilon)$ such that for $\theta > \theta_0(\varepsilon)$*

$$|\mu_\theta|_D < \varepsilon, \qquad \mu_\theta = \mu_\theta^+ - \mu_\theta^-. \tag{1.31}$$

5. *If a basis $\{h^j, j = 1, \dots, m\}$ is proper, then for each $\varepsilon > 0$ there exists $\theta_0(\varepsilon)$ such that*

$$\|\mu_\theta\|_D < \varepsilon \quad \text{when } \theta > \theta_0(\varepsilon), \qquad \mu_\theta = \mu_\theta^+ - \mu_\theta^-. \tag{1.32}$$

Thus, a set of finitely-supported measures $\{\mu_\theta\}$ that approximates μ consists, according to part 3 of the proposition, of measures "almost" orthogonal to D, while parts 4 and 5 guarantee that such "almost" orthogonality holds uniformly on D.

For the case of separated variables, we shall construct approximating measures that are exactly, not "almost," orthogonal to D.

PROOF. Consider all possible finite partitions of the basis sets $X_1, \dots, X_N$ into Borel subsets. Let $\theta_s = (U_s^1, \dots, U_s^{r_{\theta_s}})$ be a partition of X_s. For indices θ we use the set $\theta = (\theta_1, \dots, \theta_N)$ of such partitions of all basis sets. On the set θ we define the partial order relation similarly to how it was done in §4 of Chapter 2. With this, $\{\theta\}$ becomes a directed set. For a given θ consider all "parallelepipeds"

$$\Delta_\theta^{\gamma_1 \dots \gamma_N} = U_1^{\gamma_1} \times \dots \times U_N^{\gamma_n}, \qquad U_s^{\gamma_s} \in \theta_s, \quad s = 1, \dots, N, \tag{1.33}$$

that form a partition of the space Y. Inside each set $U_s^{\gamma_s}$ choose (arbitrarily) a point $x_s^{\gamma_s}$ and consider the system of points in Y

$$x_\theta^{\gamma_1 \dots \gamma_N} = x_1^{\gamma_1} \times x_2^{\gamma_2} \times \dots \times x_N^{\gamma_N} = (x_1^{\gamma_1}, x_2^{\gamma_2}, \dots, x_N^{\gamma_N}). \tag{1.34}$$

At every point $x_\theta^{\gamma_1 \dots \gamma_N}$ insert an atom μ_θ^+ and an atom μ_θ^-:

$$\mu_\theta^+(x_\theta^{\gamma_1 \dots \gamma_N}) = \mu^+(\Delta_\theta^{\gamma_1 \dots \gamma_N}), \qquad \mu_\theta^- = \mu^-(\Delta_\theta^{\gamma_1 \dots \gamma_N}),$$

so that $\|\mu_\theta^+\| + \|\mu_\theta^-\| = \|\mu^+\| + \|\mu^-\| = \|\mu\| = 1$. Also, it is clear that $\{\mu_\theta^+\}$ converges to μ^+ in the weak $(\star)$ topology while $\{\mu_\theta^-\}$ converges weak $(\star)$ to μ^-.

Relations (1.16) for $\mu \in D^\perp$ can be rewritten as follows:

$$\pi_{(j)} \circ [h^j \mu^+] = \pi_{(j)} \circ [h^j \mu^-], \qquad j = 1, \dots, m. \tag{1.35}$$

Let $(j) = (i_1, \dots, i_n)$. For a given θ consider all parallelepipeds $\Delta_\theta^{\gamma_1 \dots \gamma_N}$ with the indices $\gamma_{i_1}^0, \dots, \gamma_{i_n}^0$ being fixed (taking certain admissible values) while the others are arbitrary. Equations (1.35) yield

$$\sum \int_{\Delta_\theta^{\gamma_1 \dots \gamma_N}} h^j \, d\mu^+ = \sum \int_{\Delta_\theta^{\gamma_1 \dots \gamma_N}} h^j \, d\mu^-, \tag{1.36}$$

where the sum is taken over all $(\gamma_1, \dots, \gamma_N)$ such that $\gamma_{i_1} = \gamma_{i_1}^0, \dots, \gamma_{i_n} = \gamma_{i_n}^0$. Equalities (1.36) hold for all choices of $\gamma_{i_1}^0, \dots, \gamma_{i_n}^0$. Define h_θ^{j+} (h_θ^{j-}) at a point $x_\theta^{\gamma_1 \dots \gamma_N}$ to be the mean value on $\Delta_\theta^{\gamma_1 \dots \gamma_N}$ of h^j with respect to the measure μ^+ (μ^-). Then (1.36) becomes

$$\sum \int_{\Delta_\theta^{\gamma_1 \dots \gamma_N}} h_\theta^{j+} \, d\mu_\theta^+ = \sum \int_{\Delta_\theta^{\gamma_1 \dots \gamma_N}} h_\theta^- \, d\mu_\theta^-, \tag{1.37}$$

and this is equivalent to (1.30). For each $\varepsilon > 0$ there exists $\theta_0(\varepsilon)$ such that for $\theta > \theta_0(\varepsilon)$ the oscillation of the function h^j on any of the parallelepipeds $\Delta_\theta^{\gamma_1 \dots \gamma_N}$ is smaller than ε. Hence

$$\begin{aligned} |h_\theta^{j+}(x_\theta^{\gamma_1 \dots \gamma_N}) - h^j(x_\theta^{\gamma_1 \dots \gamma_N})| &< \varepsilon; \\ |h_\theta^{j-}(x_\theta^{\gamma_1 \dots \gamma_N}) - h^j(x_\theta^{\gamma_1 \dots \gamma_N})| &< \varepsilon. \end{aligned} \tag{1.38}$$

So far, the functions h_θ^{j+} and h_θ^{j-} are only defined at the points $x_\theta^{\gamma_1 \dots \gamma_N}$. We can extend them continuously in such a way that at every point x the inequalities

$$|h_\theta^{j+}(x) - h^j(x)| < \varepsilon, \qquad |h_\theta^{j-}(x) - h^j(x)| < \varepsilon$$

hold. Now it remains to prove parts 4 and 5. For a function $w \in D$ (of the form (4.7)) we have

$$|\mu_\theta(w)| = \left|\sum_{j=1}^{m}\sum_{\gamma_1\ldots\gamma_N} h^j\left(x_\theta^{\gamma_1\ldots\gamma_N}\right) g_j\left(\pi_{(j)}x_\theta^{\gamma_1\ldots\gamma_N}\right)\mu_\theta\left(x_\theta^{\gamma_1\ldots\gamma_N}\right)\right|$$

$$= \left|\sum_{j=1}^{m}\sum{}''\sum{}' h^j\left(x_\theta^{\gamma_1\ldots\gamma_N}\right) g_j\left(\pi_{(j)}x_\theta^{\gamma_1\ldots\gamma_N}\right)\left[\mu_\theta^+\left(x_\theta^{\gamma_1\ldots\gamma_N}\right) - \mu_\theta^-\left(x_\theta^{\gamma_1\ldots\gamma_N}\right)\right]\right|.$$

Here, $\sum'$ is taken over all $(\gamma_1,\ldots,\gamma_N)$ with fixed $(\gamma_{i_1}^0,\ldots,\gamma_{i_N}^0)$, whereas $\sum''$ is taken over all $(\gamma_{i_1}^0,\ldots,\gamma_{i_n}^0)$. Continuing the calculation and using (1.35)–(1.38), we obtain

(1.39)

$$= \left|\sum_{j=1}^{m}\sum{}'' g_j\left(\pi_{(j)}x_\theta^{\gamma_1\ldots\gamma_N}\right)\sum{}' h^j\left(x_\theta^{\gamma_1\ldots\gamma_N}\right)\left[\mu_\theta^+\left(x_\theta^{\gamma_1\ldots\gamma_N}\right) - \mu_\theta^-\left(x_\theta^{\gamma_1\ldots\gamma_N}\right)\right]\right|$$

$$= \left|\sum_{j=1}^{m}\sum{}'' g_j\left(\pi_{(j)}x_\theta^{\gamma_1\ldots\gamma_N}\right)\sum{}' h^j\left(x_\theta^{\gamma_1\ldots\gamma_N}\right)\left[\mu_\theta^+\left(x_\theta^{\gamma_1\ldots\gamma_N}\right) - \mu_\theta^-\left(x_\theta^{\gamma_1\ldots\gamma_N}\right)\right]\right.$$
$$\left. - \sum_{j=1}^{m}\sum{}'' g_j\left(\pi_{(j)}x_\theta^{\gamma_1\ldots\gamma_N}\right)\sum{}' \left[h^{j+}\left(x_\theta^{\gamma_1\ldots\gamma_N}\right)\mu_\theta^+\left(x_\theta^{\gamma_1\ldots\gamma_N}\right) - h_\theta^{j-}\left(x_\theta^{\gamma_1\ldots\gamma_N}\right)\mu_\theta^-\left(x_\theta^{\gamma_1\ldots\gamma_N}\right)\right]\right|$$

$$\leq \sum_{j=1}^{m}\|g_j\|\cdot\varepsilon\left(\|\mu_\theta^+\| + \|\mu_\theta^-\|\right) \leq C|w|\varepsilon.$$

The inequality (1.39) means that $|\mu_\theta|_D < C\varepsilon$. If the basis system is proper, (1.39) also implies the estimate $\|\mu_\theta\|_D < C_1\varepsilon$. The proof of Proposition 1.7 is now complete.

REMARK 1.8. Clearly, one can assume that $\|\mu_\theta\| = \|\mu_\theta^+ - \mu_\theta^-\| = 1$. Indeed, on one side $\|\mu_\theta\| \leq \|\mu_\theta^+\| + \|\mu_\theta^-\| = 1$. On the other side, $\|\mu\| \leq \underline{\lim}\|\mu_\theta\|$ and therefore for an arbitrary $\varepsilon > 0$ there exists $\theta_0(\varepsilon)$ such that $\theta > \theta_0 \Rightarrow \|\mu_\theta\| > 1 - \varepsilon$. Hence, instead of $\{\mu_\theta\}$ we could use $\tilde{\mu}_\theta = \mu_\theta/\|\mu_\theta\|$. (In general, the equality $\mu_\theta = \mu_\theta^+ - \mu_\theta^-$ is not the Jordan decomposition of the measure μ_θ.)

A reminder on projections. Let B be a Banach space and $E \subset B$ a subspace of B. A continuous linear operator $P : B \to E$ is called a projection (of B onto E) if

$$P(B) = E, \qquad P^2 = P \circ P = P \tag{1.40}$$

(idempotence). If we set $R = I - P$, where I is, as usual, the identity operator, then R is also a projection, and if we set $\mathcal{E} = R(B)$, then

$$B = E \dot{+} \mathcal{E} \tag{1.41}$$

($\dot{+}$ means that every $x \in B$ can be represented in the form $x = x' + x''$, where $x' \in E$, $x'' \in \mathcal{E}$, and such a representation is unique). Moreover,

$$E = \{x : Rx = \theta\}, \qquad \mathcal{E} = \{x : Px = 0\}; \tag{1.42}$$

hence E and $\mathcal{E}$ are closed subspaces. Conversely, if (1.41) holds with closed subspaces E and $\mathcal{E}$, then $P : B \to E$ ($Px = P(x' + x'') = x'$) and $R : B \to \mathcal{E}$ ($Rx = R(x' + x'') = x''$) are projections. Thus, for existence of a projection onto a subspace it is necessary and sufficient that the subspace be closed and have a closed complement (with respect to the sum (1.41)).

The projections P and $R = I - P$ are called complementary. The projection R is also called residual with respect to P.

LEMMA 1.9. *Let $P_i : B \to E_i$, $i = 1, \dots, n$, be projections, commuting with each other: $P_iP_j = P_jP_i$, $1 \le i \le j \le n$. Then the operator*

$$P = \sum_i P_i - \sum_{i<j} P_iP_j + \sum_{i<j<k} P_iP_jP_k + \dots + (-1)^{n-1}P_1 \dots P_n \tag{1.43}$$

is a projection of B onto

$$E_1 + E_2 + \dots + E_n. \tag{1.44}$$

If R_i denotes the complementary projections for P_i, $i = 1, \dots, n$, then

$$R = R_1R_2 \dots R_n \tag{1.45}$$

is the complementary projection for P.

Lemma 1.9 is commonly used for $n = 2$ (cf., e.g., [**40,** p. 518]); the general case follows by induction. It allows us to characterize the subspace D in our problems by some finite difference-like conditions. Let us illustrate this by a simple example.

EXAMPLE. In the space $C(Y)$ consider a subspace D that consists of functions

$$g_1(x_2, \dots, x_N) + g_2(x_1, x_3, \dots, x_N) + \dots + g_N(x_1, \dots, x_{N-1}). \tag{1.46}$$

Let D_j, $j = 1, \dots, N$, consist of all functions of the form

$$g_j(x_1, \dots, x_{j-1}, x_{j+1}, \dots, x_N). \tag{1.47}$$

Fix a point $(x_1^0, \dots, x_N^0)$ and consider an operator $P_j : C(Y) \to D_j$ defined by

$$P_j : f(x_1, \dots, x_n) \to f(x_1, \dots, x_j^0, \dots, x_n). \tag{1.48}$$

P_j is a projection of $C(Y)$ onto D_j. The complementary projection $R_j = I - P_j$ is defined by

$$f(x_1, \dots, x_N) \to f(x_1, \dots, x_N) - f(x_1, \dots, x_j^0, \dots, x_N) \tag{1.49}$$

and D_j consists of all $f \in C(Y)$ for which the finite difference (1.49) vanishes. The operators P_j, $j = 1, \dots, N$, commute, so that D consists of all functions $f \in C(Y)$ for which

$$R_1R_2 \dots R_N(f) = 0. \tag{1.50}$$

Moreover, the operator $R_1 \dots R_N$ turns out to be a finite difference of order N.

Separated variables. Projections on D_j. Consider the subspace D (1.11). Let D_j be the subspace

$$D_j = H^j(\widetilde{Y}_j) \otimes C(Y_j), \qquad j = 1, \dots, m. \tag{1.51}$$

Take an arbitrary basis $h_j^\delta(\tilde{y}_j)$, $\delta = 1, \dots, n_j$, for $H^j(\widetilde{Y}_j)$. The subspace D_j consists of functions

$$\sum_{\delta=1}^{n_j} h_j^\delta(\tilde{y}_j^\delta) g_j^\delta(y_j), \qquad g_j^\delta \in C(Y_j). \tag{1.52}$$

Denote by λ_j^δ linear functionals that are biorthogonal to the basis $\{h_j^\delta\}$. A functional λ_j^δ is represented by a measure on $\widetilde{Y}_j$. Consider a linear operator $P = P_j : C(Y) \to D_j$ defined by

$$f \in C(Y) \mapsto \sum_{\delta=1}^{n_j} \lambda_j^\delta \left[f\left(\tilde{y}_j, \underbrace{y_j}\right)\right] h_j^\delta(\tilde{y}_j), \tag{1.53}$$

where $\underbrace{\quad}$ indicates the variables that are assumed constant (parameters) under the action of the functional λ_j^δ.

LEMMA 1.10. *The operator $P = P_j$ is a projection of $C(Y)$ onto D_j. Moreover, for any $f \in C(Y)$ the function $F = Pf$ satisfies the interpolational conditions*

$$\lambda_j^\delta \left[f\left(\tilde{y}_j, \underbrace{y_j}\right) - F\left(\tilde{y}_j, \underbrace{y_j}\right)\right] = 0 \tag{1.54}$$

and is the unique function in D_j satisfying (1.54). *The system of the basis functions $\{h_j^\delta\}$, $\delta = 1, \dots, n_j$, is proper in $C(Y)$.*

PROOF. Let Φ be an arbitrary function in D_j,

$$\Phi(\tilde{y}_j, y_j) = \sum_{\delta=1}^{n_j} h_j^\delta\left(\widetilde{Y}_j\right) g^\delta(y_j), \qquad g^\delta \in C(Y_j). \tag{1.55}$$

We have

$$\begin{aligned} \lambda_j^{\delta_0}\left[\Phi\left(\tilde{y}_j, \underbrace{y_j}\right)\right] &= \sum_{\delta=1}^{n_j} \lambda_j^{\delta_0}\left[h_j^\delta(\tilde{y}_j) g^\delta(y_j)\right] \\ &= \sum_{\delta=1}^{n_j} g^\delta(y_j) \lambda_j^{\delta_0}\left(h_j^\delta(\tilde{y}_j)\right) = g^{\delta_0}(y_j). \end{aligned} \tag{1.56}$$

From (1.56) it follows that the representation (1.55) of any function $\Phi \in D_j$ is unique and the coefficients $g^\delta(y_j)$ at $h_j^\delta(\tilde{y}_j)$ are found by the formula (1.56). Therefore, for an arbitrary $\Phi \in D_j$ we have

$$P\Phi = \sum_{\delta-1}^{n_j} \lambda_j^\delta\left[\Phi\left(\tilde{y}_j, \underbrace{y_j}\right)\right] h_j^\delta(\tilde{y}_j) = \sum_1^{n_j} h_j^\delta(\tilde{y}_j) g_\delta(y_j) = \Phi.$$

Thus, $P[C(Y)] = D_j$ and $P^2 = P$. Combining (1.53) and (1.56) we get (1.54). Since the representation (1.55) is unique, Pf is the unique function in D_j that satisfies the interpolational conditions (1.54). From the uniqueness of the representation (1.55) and formulas (1.56) it also follows that the system $\{h_j^\delta\}$ is proper.

Commutativity of projections. Let (j) and (k) be two sets of indices (1.5). Assume that they may overlap but cannot completely contain one another. Therefore, the same holds for the complementary sets $(\tilde{j})$ and $(\tilde{k})$. Denote

$$(s) = (\tilde{j}) \cap (\tilde{k}), \qquad (\sigma) = (j) \cap (k), \qquad (j)\backslash(\sigma) = (j_1),$$
$$(\tilde{j})\backslash(s) = (j_2), \qquad (k)\backslash(\sigma) = (k_1), \qquad (\tilde{k})\backslash(s) = (k_2). \tag{1.57}$$

A complete set of indices N can now be represented as follows:

$$(N) = (s) \cup (j_2) \cup (\sigma) \cup (j_1) = (s) \cup (k_2) \cup (\sigma) \cup (k_1). \tag{1.58}$$

It is easy to see that

$$(j_2) = (k_1), \qquad (j_1) = (k_2). \tag{1.59}$$

Indeed, $(k_2) \cap (j_2) = \emptyset$; otherwise, (s) could not be the intersection of $(\tilde{j})$ with $(\tilde{k})$. Therefore, $(k_2) \subset (j)$. If there existed an index $i \in (\sigma) \cap (k_2)$, then $i \in (k)$, which cannot happen because $(k_2) \subset (\tilde{k})$. So, $(k_2) \subset (j)\backslash(\sigma) = (j_1)$. Yet, along the same lines $(j_1) \cap (k_1) = \emptyset$ and $(j_1) \subset (\tilde{k})\backslash(s) = (k_2)$. We have proved one of the equalities (1.59); the second is proved similarly.

Consider in $C(Y)$ the subspaces

$$D = D_j = H(\widetilde{Y}_{(j)}) \otimes C(Y_{(j)});$$
$$D = D_k = H(\widetilde{Y}_{(k)}) \otimes C(Y_{(k)}). \tag{1.60}$$

The subspaces H are assumed to have the following structure:

$$H(\widetilde{Y}_{(j)}) = H(Y_{(\tilde{j})}) = H(Y_{(s)}) \otimes H(Y_{(j_2)});$$
$$h(\widetilde{Y}_{(k)}) = H(Y_{(\tilde{k})}) = H(Y_{(s)}) \otimes H(Y_{(k_2)}), \tag{1.61}$$

where all H are finite-dimensional subspaces of continuous functions of the variables in parentheses. Consider some bases in those subspaces:

$$\{c^\alpha(y_{(s)})\}, \qquad \alpha = 1, \dots, \ell \text{ – a basis in } H(Y_{(s)}); \tag{1.62$_1$}$$

$$\{d^\beta(y_{(j_2)})\}, \qquad \beta = 1, \dots, n \text{ – a basis in } H(Y_{(j_2)}); \tag{1.62$_2$}$$

$$\{e^\gamma(y_{(k_2)})\}, \qquad \gamma = 1, \dots, r \text{ – a basis in } H(Y_{(k_2)}). \tag{1.62$_3$}$$

The systems

$$\{c^\alpha(y_{(s)}) \otimes d^\beta(y_{(j_2)})\} \quad \text{and} \quad \{c^\alpha(y_{(s)}) \otimes e^\gamma(y_{(k_2)})\}$$

are bases in $H(\widetilde{Y}_{(j)})$ and $H(\widetilde{Y}_{(k)})$, respectively. Also, let $\{\lambda^\alpha\}$, $\alpha = 1, \dots, \ell$, $\{\mu^\beta\}$, $\beta = 1, \dots, n$, $\{\nu^\gamma\}$, $\gamma = 1, \dots, r$, be the systems of linear functionals biorthogonal to the bases (1.62$_1$), (1.62$_2$), and (1.62$_3$), respectively. The functionals λ^α are represented by measures on $Y_{(s)}$, μ^β are represented by measures on $Y_{(j_2)}$, ν^γ are represented by measures on $Y_{(k_2)}$. Measures $\{\lambda^\alpha \otimes \mu^\beta\}$ form a system biorthogonal to the basis chosen in $H(\widetilde{Y}_{(j)})$, while $\{\lambda^\alpha \otimes \nu^\gamma\}$ form a system biorthogonal to the basis in $H(\widetilde{Y}_{(k)})$. Consider projections $P : C(Y) \to D = D_j$ and $\mathcal{P} : C(Y) \to D = D_k$ defined by

$$
\begin{aligned}
&P : f \in C(Y) \mapsto \sum_{\alpha,\beta} \lambda^\alpha \otimes \mu^\beta \left[f\left(y_{(s)}, y_{(j_2)}, \underbrace{y_{(j)}} \right) \right] c^\alpha\left(y_{(s)}\right) d^\beta\left(y_{(j_2)}\right);\\
(1.63)\qquad &\mathcal{P} : f \in C(Y) \mapsto \sum_{\delta,\gamma} \lambda^\delta \otimes \nu^\gamma \left[f\left(y_{(s)}, y_{(k_2)}, \underbrace{y_{(k)}} \right) \right] c^\delta\left(y_{(s)}\right) e^\gamma\left(y_{(k_2)}\right).
\end{aligned}
$$

LEMMA 1.11. *Whenever $H(\widetilde{Y}_{(j)})$ and $H(\widetilde{Y}_{(k)})$ have the structure* (1.61), *the projections P and $\mathcal{P}$ commute*

First of all, note that we can permute actions of the linear functionals λ^α, μ^β, ν^γ. To see that, it suffices to take their Riesz representations and apply the Fubini theorem. Set

$$
(1.64)\qquad \varphi^{\alpha\beta}\left(y_{(j)}\right) = \varphi^{\alpha\beta}\left(y_{(\sigma)}, y_{(k_2)}\right) = \lambda^\alpha \otimes \mu^\beta \left[f\left(y_{(s)}, y_{(j_2)}, \underbrace{y_{(j)}} \right) \right].
$$

Using biorthogonality of $\{\lambda^\alpha\}$ and $\{c^\alpha\}$, we have
(1.65)

$$
\begin{aligned}
\mathcal{P}(Pf) &= \sum_{\delta,\gamma} \lambda^\delta \otimes \nu^\gamma \left\{ \sum_{\alpha,\beta} \varphi^{\alpha\beta}(y_{(j)}) \otimes c^\alpha(y_{(s)}) \otimes d^\beta(y_{(j_2)}) \right\} \otimes c^\delta(y_{(s)}) \otimes e^\gamma(y_{(k_2)})\\
&= \sum_{\tilde{\sigma},\gamma,\beta} \nu^\gamma \left\{ \varphi^{\delta\beta}\left(\underbrace{y_{(\sigma)}}, y_{(k_2)} \right) \right\} \otimes c^\delta\left(y_{(s)}\right) \otimes d^\beta\left(y_{(j_2)}\right) \otimes e^\gamma\left(y_{(k_2)}\right).
\end{aligned}
$$

But

$$
(1.66)\qquad \nu^\gamma \left\{ \varphi^{\delta\beta}\left(\underbrace{y_{(\sigma)}}, y_{(k_2)} \right) \right\} = \nu^\gamma \otimes \lambda^\delta \otimes \mu^\beta \left[f\left(\underbrace{y_{(\sigma)}}, y_{(s)}, y_{(j_2)}, y_{(k_2)} \right) \right].
$$

Calculating $P(\mathcal{P}f)$ and using (1.65)–(1.66) (with an opposite order of action of functionals), we arrive at the same result. Since actions of functionals commute, $P\mathcal{P} = \mathcal{P}P$.

Totally separated variables. Projections onto D. Consider the situation (1.13) with totally separated variables. If $\{h_i^\alpha(x_i)\}$, $\alpha = 1, \dots, n_i$, is a basis of $H^i(X_i)$, $i = 1, \dots, N$, then bases $H^j(\widetilde{Y}_j)$ are formed by the functions $h_j^\delta(\tilde{y}_j)$ that are given by products (1.24). If $\{\lambda_i^\alpha\}$, $\alpha = 1, \dots, n_i$, are linear functionals in $C(X_i)^\star$ biorthogonal to the basis $\{h_i^\alpha\}$, $(\bar{j}) = (i_1, \dots, i_{k_{\bar{j}}})$, then all possible tensor products

$$
(1.67)\qquad L^\delta = L_j^\delta = \lambda_{i_1}^\alpha \otimes \cdots \otimes \lambda_{i_{k_{\bar{j}}}}^\gamma
$$

give the system of linear functionals in $C(\widetilde{Y}_j)^\star$ which is biorthogonal to the basis (1.24). As above, set

$$
(1.68)\qquad D_j H^j\left(\widetilde{Y}_j\right) \otimes C\left(Y_j\right), \qquad D = \sum_{j=1}^m D_j,
$$

and define $P_j : C(Y) \to D_j$, $j = 1, \ldots, m$, by formulas similar to (1.53):

$$f \in C(Y) \to P_j(f) = \sum_{\delta} L_j^{\delta} \left[f \left(\tilde{y}_j, \underbrace{y_j} \right) \right] h_j^{\delta} \left(\tilde{y}_j \right). \tag{1.69}$$

THEOREM 1.12. *P_j is a projection of $C(Y)$ onto D_j, $j = 1, \ldots, m$. Any two operators P_j commute. The operator P defined by* (1.43) *is a projection of $C(Y)$ onto D. If the R_j are complementary projections for the P_j, then the operator R defined by* (1.45) *is the complementary projection for P. For any $f \in C(Y)$ the function $F = Pf$ satisfies the interpolational conditions*

$$L_j^{\delta} \left[f \left(\tilde{y}_j, \underbrace{y_j} \right) - F \left(\tilde{y}_j, \underbrace{y_j} \right) \right] = 0, \tag{1.70}$$

and is the unique function in D satisfying (1.70). *The system of basis functions $\left\{ h_j^{\delta} \left(\tilde{y}_j \right), j = 1, \ldots, m, \delta = 1, \ldots, N_j \right\}$ is proper in $C(\widetilde{Y}_j)$.*

PROOF. The fact that P_j is a projection from $C(Y)$ onto D_j is proved in Lemma 1.10. Since the variables are totally separated for each pair P_j and P_k, the assumptions of Lemma 1.11 hold and therefore P_j and P_k commute. The operator P defined in (1.43) is a projection onto D by Lemma 1.9, whereas R in (1.45) is the projection complementary to P. To establish the interpolation identities (1.70), fix an index j. The equalities (1.54) of Lemma 1.10 can be rewritten in the following form: for each $f \in C(Y)$ we have $L_j^{\delta}[\Phi] = 0$, where we set $\Phi = f - P_j f = R_j f$. Now, for all $f \in C(Y)$ let $\Phi = f - Pf = Rf$. In view of (1.45),

$$\begin{aligned} \Phi &= R_j \left[R_1 \ldots R_{j-1} R_{j+1} \ldots R_m f \right] = R_j^2 \left[R_1 \ldots R_{j-1} R_{j+1} \ldots R_m f \right] \\ &= R_j \left[R_1 \ldots R_{j-1} R_j R_{j+1} \ldots R_m f \right] = R_j [Rf], \end{aligned}$$

so either $L_j^{\delta}[Rf] = 0$ or $L_j^{\delta}[f - Pf] = 0$. We have proved (1.70) by using idempotency of the projection R_j and commutativity of the projections under consideration. To prove the remaining statements, turn to the identity (1.43) and rewrite it as follows:

$$\begin{aligned} P = P_1 &+ P_2 (I - P_1) + P_3 (I - P_1 - P_2 + P_1 P_2) \\ &+ \cdots + P_m [I - P_1 - \cdots - P_{m-1} + P_1 P_2 + \cdots + P_{m-2} P_{m-1} - P_1 P_2 P_3 \\ &\qquad - \cdots - P_{m-3} P_{m-2} P_{m-1} + \cdots + (-1)^{m-2} P_1 \ldots P_{m-1}]. \end{aligned}$$

Using (1.45) for complementary projections, we can rewrite the latter identity in the following form:

$$P = P_1 + P_2 R_1 + P_3 R_1 R_2 + \cdots + P_m R_1 R_2 \ldots R_{m-1}. \tag{1.71}$$

Therefore, for all $f \in C(Y)$ we have

$$\begin{aligned} w &= Pf = w_1 + w_2 + \cdots + w_m, \quad \text{where} \\ w_1 &= P_1 f \in D_1, \\ w_2 &= P_2 R_1 f \in D_2, \ldots, w_j = P_j R_1 \ldots R_{j-1} f \in D_j, \ldots \\ w_m &= P_m R_1 \ldots R_{m-1} f \in D_m. \end{aligned} \tag{1.72}$$

Whenever $f = w \in D$, instead of f everywhere in (1.72) we can put $w = Pw$.

Repeating the arguments in the derivation of the interpolation identities (1.70) we obtain, in view of the structure of terms w_j in (1.72), the following formula:

$$\begin{aligned} &L_1^\delta(w_2) = L_1^\delta(w_3) = \cdots = L_1^\delta(w_m) = 0, \qquad \delta = 1,\dots,N_1;\\ &L_2^\delta(w_3) = L_2^\delta(w_4) = \cdots = L_2^\delta(w_m) = 0, \qquad \delta = 1,\dots,N_2;\\ &L_{m-1}^\delta(w_m) = 0, \qquad \delta = 1,\dots,N_{m-1}. \end{aligned} \tag{1.73}$$

Now let $w^\star \in D$ be an element such that

$$L_j^\delta(w^\star) = L_j^\delta(f), \qquad j = 1,\dots,m,\ \delta = 1,\dots,N_j, \tag{1.74}$$

and let

$$w^\star = w_1^\star + w_2^\star + \cdots + w_m^\star \tag{1.75}$$

be its (1.72) decomposition into D_j-components. Since $L_1^\delta(f) = L_1^\delta(w^\star)$ and the values of the functionals L_1^δ completely determine an element in D_1, $w_1 = w_1^\star$. Also,

$$L_2^\delta(f) = L_2^\delta(w_1) + L_2^\delta(w_2) = L_2^\delta(w^\star) = L_2^\delta(w_1^\star) + L_2^\delta(w_2^\star),$$

and so $L_2^\delta(w^\star) = L_2^\delta(w_2)$. However, the values of the functional L_2^δ uniquely determine elements in D_2; hence $w_2^\star = w_2$. Continuing in this fashion with the functionals L_3^δ, we see that $w_3^\star = w_3$, etc. (In all arguments with L_j^δ the index δ runs over all the values $1,\dots,N_j$.) Thus, $w = Pf = w^\star$ and we have proved that Pf is the unique element in D satisfying the interpolational conditions (1.70).

Now, let us show that a combined system of bases $\{h_j^\delta\}$ is proper. From (1.72) it follows, in particular, that

$$\|w_j\| \le C\|w\|, \qquad j = 1,\dots,m, \tag{1.76}$$

with an appropriate constant C. Since for all j the system $\{h_j^\delta\}$, $\delta = 1,\dots,N_j$, is proper (Lemma 1.10), (1.76) yields that the combined system $\{h_j^\delta\}$, $j = 1,\dots,m$, $\delta = 1,\dots,N_j$, is also proper. The latter could also be proven by a simpler argument based on Proposition 1.1: if there is a projection from $C(Y)$ onto D, then D is a closed subspace and hence the system of basis functions is proper.

A special case: point-evaluation functionals. For the linear functionals in the formulas considered above one can take point-evaluations. A linear functional relating to a function f its value at a fixed point x in its domain of definition will be, for brevity, denoted by $\hat{x}$, or $x^\wedge$. So,

$$\hat{x}(f) = f(x). \tag{1.77}$$

Such a functional is represented by the delta-measure δ_x supported at the point x.

The following two facts are almost obvious and well-known.

LEMMA 1.13. *Let $\varphi_1(x),\dots,\varphi_n(x)$ be linearly independent functions on a set X. Then there exist n points $x_i \in x$, $i = 1,\dots,n$, such that*

$$\begin{vmatrix} \varphi_1(x_1) & \cdots & \varphi_1(x_n)\\ \varphi_2(x_1) & \cdots & \varphi_2(x_n)\\ \vdots & & \vdots\\ \varphi_n(x_1) & \cdots & \varphi_n(x_n) \end{vmatrix} \ne 0. \tag{1.78}$$

PROOF. Since $\varphi_1(x) \not\equiv 0$, take x_1 with $\varphi_1(x_1) \neq 0$. Consider the determinant

$$\Delta_2(x) = \begin{vmatrix} \varphi_1(x_1) & \varphi_1(x) \\ \varphi_2(x_1) & \varphi_2(x) \end{vmatrix}.$$

If $\Delta_2(x) \equiv 0$, $\varphi_2(x)$ would be a scalar multiple of $\varphi_1(x)$, which cannot happen. Hence, there exists a point x_2 with $\Delta_2(x_2) \neq 0$. Consider the determinant

$$\Delta_3(x) = \left| \begin{array}{cc|c} \multicolumn{2}{c|}{\Delta_2(x_2)} & \begin{array}{c}\varphi_1(x) \\ \varphi_2(x)\end{array} \\ \hline \varphi_3(x_1) & \varphi_3(x_2) & \varphi_3(x) \end{array} \right|$$

and, similarly, find a point x_3 at which $\Delta_3(x_3) \neq 0$, etc.

COROLLARY 1.14. *Let H be an n-dimensional space of functions defined on a set X. There exist n points $x_i \in X$, $i = 1, \dots, n$, such that the linear functionals $\widehat{x}_1, \dots, \widehat{x}_n$ are linearly independent over H and there exists a basis in H biorthogonal to $\{\widehat{x}_i,\ i = 1, \dots, n\}$.*

PROOF. The points where (1.78) holds satisfy the corollary. A basis biorthogonal to $\{\widehat{x}_i\}$ is found among linear combinations of a given basis $\{\varphi_i(x)\}$ by solving a system of equations expressing biorthogonality.

COROLLARY 1.15. *The statement of Lemma 1.10 holds if $\lambda_j^\alpha = \widehat{\tilde{y}_j^\alpha}$, $\alpha = 1, \dots, n_j$, $\tilde{y}_j^\alpha \in \widetilde{Y}_j$, while $\{h_j^\alpha(\tilde{y}_j)\}$ is a basis in $H^j(\widetilde{Y}_j)$ biorthogonal to linearly independent functionals $\tilde{y}_j^\alpha$. The statement of Theorem 1.12 also holds when $\lambda_i^\alpha = \widehat{x}_i^\alpha$ are linearly independent functionals, $x_i^\alpha \in X_i$, and $\{h_i^\alpha(x_i)\}$, $i = 1, \dots, N$, $\alpha = 1, \dots, n_i$, is a basis in $H^i(X_i)$ biorthogonal to it.*

Thus, in (1.54) and (1.70) we are now dealing with the ordinary point interpolation.

The nature of continuity and smoothness of projections on D. Let λ_j^α be a linear functional on $C(\widetilde{Y}_j)$. Then, the function

$$\lambda_j^\alpha \left[f\left(\tilde{y}_j, \underbrace{y_j}\right) \right] \tag{1.79}$$

obviously has continuity with respect to the variables y_j (the parameter, with respect to the action of λ_j^α) that is not worse than the original function. Thus, under projections (1.53) or (1.72), characteristics (moduli) of continuity are inherited by properties of a function f and basis functions $\{h_j^\alpha\}$. The same can be said about smoothness properties whenever f and $\{h_j^\alpha\}$ have any.

PROOF OF PROPOSITION 1.6 ON "DOMINO"-MEASURES. As noted above, necessity follows from Lemma 1.5. Let us prove sufficiency. Let x_i^α, $\alpha = 1, \dots, n_i$, $i = 1, \dots, N$, be points in the basis sets X_i such that the functionals $\widehat{x}_i^\alpha$ are linearly independent over $H^i(X_i)$, and let $\{h_i^\alpha(x_i)\}$ be a biorthogonal basis in $H^i(X_i)$. Let the projection P from $C(Y)$ onto D be defined by (1.45) and (1.69), where now $\lambda_i^\alpha = \widehat{x}_i^\alpha$. Construct the measure μ_i on X_i $(i = 1, \dots, N)$ as follows:

$$\mu_i = \sum_{\alpha=1}^{n_i} a_i^\alpha \widehat{x}_i^\alpha + a_i^0 \widehat{x}_i^0 \in H^i(X_i)^\perp, \tag{1.80}$$

where x_i^0 is an arbitrary point in X_i. This can always be done so that $a_i^0 \neq 0$. Then the measure $\mu = \mu_1 \otimes \mu_2 \otimes \cdots \otimes \mu_N$ is a domino-measure.

Apply Theorem 1.12, setting

$$L_j^\delta = \widehat{x}_j^\delta, \qquad \delta = 1, \dots, n_j; \qquad h_j^\delta = h_j^\delta(x_j).$$

Let the function $f \in C(Y)$ be orthogonal to μ. But in view of Lemma 1.5 the function $F = Pf$ is always orthogonal to μ. Therefore, $\Phi = f - Pf \perp \mu$. We have

$$(1.81) \quad 0 = \mu(\Phi) = \sum a_1^\alpha \dots a_N^\alpha \Phi(x_1^\alpha \times x_2^\beta \times \cdots \times x_N^\gamma) + a_1^0 a_2^0 \dots a_N^0 \Phi(x_1^0 \dots x_N^0),$$

where the summation is extended over all admissible sets of indices $(\alpha, \beta, \dots, \gamma) \neq (0, 0, \dots)$. The interpolation conditions (1.70) now imply that Φ vanishes on each of the hyperplanes $x_1 = x_1^\alpha$, $\alpha = 1, \dots, n_1$; $x_2 = x_2^\beta$, $\beta = 1, \dots, n_2; \dots; x_N = x_N^\gamma$, $\gamma = 1, \dots, n_N$. Therefore, the first sum in (1.81) vanishes. Hence, $\Phi\left(x_1^0, \dots, x_N^0\right) = 0$ at every point $\left(x_1^0, \dots, x_N^0\right)$ and $f - Pf \equiv 0$.

COROLLARY 1.16. *Under the assumptions of Proposition* 1.6 *any measure* $\mu \in D^\perp$ *is a weak* ($\star$) *limit of a net* $\{\mu_\theta\}$ *of measures that are (finite) linear combinations of "domino"-measures.*

The statement follows from general functional-analytic considerations (e.g., [**40**], or [**118**]). Unfortunately, general considerations cannot guarantee that a net $\{\mu_\theta\}$ can be chosen so that $\|\mu_\theta\| \to \|\mu\|$ (cf., e.g., [**118**]), which is important for applications.

Interpolation and norms on subsets. Functions in the unit ball of the subspace D need not, generally speaking, be equicontinuous. However, as we shall show below, there still exist some substitutes for this obvious (when compared with the case of separated variables) loss. Recall that $F|_E$ denote the restriction of a function F to a set E.

LEMMA 1.17. *Let a subspace* D_j *have the form* (1.51), *and let* R *be an arbitrary subset in* $Y_{(j)}$. *There exists a linear operator* $T : C(Y) \to C(Y)$ *with the following properties*:

$$(1.82) \quad \begin{aligned} &1. \quad (Tf - f)\Big|_{R \times \widetilde{Y}_j} = 0; \\ &2. \quad \|Tf\| \leq \|f\|_{R \times \widetilde{Y}_j}; \\ &3. \quad T(D_j) \subset D_j. \end{aligned}$$

PROOF. Let $R = \left\{y_j^1, \dots, y_j^\nu\right\}$. Take disjoint neighborhoods V^k, $k = 1, \dots, \nu$, of the points $y_j^k \in V^k$. Construct functions $\Phi^k(y_i) \subset C(Y_j)$ with the following properties:

$$(1.83) \qquad \Phi^k\left(y_j^k\right) = 1, \qquad \Phi^k(y_j) = 0, \quad y_j \in Y_j \backslash V^k, \quad \left\|\Phi^k\right\| = 1,$$

and define an operator T by setting

$$(1.84) \qquad f \in C(Y) \to Tf = \sum_{k=1}^{\nu} \Phi^k(y_j) \otimes f\left(\tilde{y}_j, y_j^k\right).$$

Let us verify that the properties (1.82) hold. If $y = \left(\tilde{y}_j, y_j^k\right)$, then according to (1.83) $Tf\left(\tilde{y}_j, y_j^k\right) = f\left(\tilde{y}_j, y_j^k\right)$ and property 1 is proved. Let us prove property 2 in (1.82). If $(\tilde{y}_i, y_j) \notin \widetilde{Y}_j \times \bigcup_k V^k$, then $\Phi^k(y_j) = 0$, $k = 1, \dots, \nu$, so that $Tf(\tilde{y}_j, y_j) = 0$. If $y_j \in V^k\left(y_j^k\right)$, then $\Phi^s(y_j) = 0$, $s \neq k$. Therefore, $|Tf(\tilde{y}_j, y_j)| = \left|\Phi^k(y_j)\right| \left|f\left(\tilde{y}_j, y_j^k\right)\right| \leq \left\|f\left(\tilde{y}_j, y_j^k\right)\right\|_{\widetilde{Y}_j} \leq \|f(\tilde{y}_j, y_j)\|_{R\times\widetilde{Y}_j}$. So, $\|Tf\| \leq \|f\|_{R\times\widetilde{Y}_j}$. Now, let $f \in D_j$, i.e.,

$$f = \sum_1^{n_j} h_j^\delta(\tilde{y}_j)\,\varphi^\delta(y_j) \quad \text{for some } \varphi^\delta(y_j) \in C(Y_j).$$

Therefore

$$f\left(\tilde{y}_j, y_j^k\right) = \sum_1^{n_j} h_j^\delta(\tilde{y}_j)\,\varphi^\delta\left(y_j^k\right) \in H^j\left(\widetilde{Y}_j\right),$$

and $Tf \in D_j$. The lemma is proved. (The operator T, generally speaking, need not be a projection.)

Now let us consider the subspace D assuming that variables are totally separated: D is defined by the formulas (1.12)–(1.13). In each basic set X_s find points $x_s^1, \dots, x_s^{n_j}$ and a corresponding basis $\{h_s^\alpha\}$, $\alpha = 1, \dots, n_s$, in H^s, so that the functionals $\widehat{x}_s^\alpha$ and $\{h_s^\alpha\}$ are biorthogonal. As in Proposition 1.7, considering partitions Θ_s of the set X_s, $\Theta_s = \left\{U_s^{\delta_s}\right\}$ and taking a point $x_s^{\alpha_s}$ in each $U_s^{\delta_s}$ we agree to include among those points the fixed points $x_s^1, \dots, x_s^{n_s}$. Clearly, when Θ_s is sufficiently large, i.e., for sufficiently fine partitions, this can be done. Similarly to Proposition 1.7, as the set of indices we consider the set of all partitions $\Theta = (\Theta_1, \dots, \Theta_s, \dots, \Theta_N)$ of the basis sets X_s.

Fix Θ and let A_s be a (finite) set of points in X_s obtained by taking a point in each $U_s^{\delta_s}$ (with the above agreement concerning inclusion of the set $A_s^\star = \{x_1^s, \dots, x_s^{n_s}\}$ into $A_s : A_s^\star \subset A_s$). Denote $A = A_1 \times \dots \times A_N$. Let us prove the following theorem on interpolation.

THEOREM 1.18. *In the structure of D, let the variables be totally separated. There exist an absolute constant M and Θ_0 such that for all $\Theta > \Theta_0$ and all $w \in D$ one can find $w' \in D$ with the following properties*:

$$w|_{A^\cdot} = w'|_{A^\cdot}, \qquad A = A_1 \times \dots \times A_N, \qquad \|w'\| \leq M\|w\|_A. \tag{1.85}$$

PROOF. Take an index j, $j = 1, \dots, m$, and let

$$(\tilde{j}) = (i_1, \dots, i_{\tilde{k}_j}), \qquad (j) = (\ell_1, \dots, \ell_{k_j}).$$

Consider the products

$$\begin{aligned} \tilde{A}_{(j)} &= A_{(\tilde{j})} = A_{i_1} \times \dots \times A_{i_{\tilde{k}_j}}, \\ A_{(j)} &= A_{\ell_1} \times \dots \times A_{\ell_{k_j}}; \\ A^\star_{(\tilde{j})} &= A^\star_{\ell_1} \times \dots \times A^\star_{\ell_{\tilde{k}_j}}, \\ A^\star_{(j)} &= A^\star_{\ell_1} \times \dots \times A^\star_{\ell_{k_j}}. \end{aligned} \tag{1.86}$$

Clearly, $A^\star_{(\bar{j})} \subset A_{(\bar{j})}$, $A^\star_{(j)} \subset A_{(j)}$, $A_{(\bar{j})} \times A_{(j)} = A$. Define projections $P_j : C(Y) \to D_j$, $j = 1, \dots, m$, according to the formulas (1.67)–(1.69), where we now set

$$\Lambda_j^\delta = \widehat{x}_{i_1}^\alpha \otimes \dots \otimes \widehat{x}_{i_{\bar{k}_j}}^\alpha = \widehat{x}_{i_1}^\alpha \times \dots \times x_{i_{\bar{k}_j}}^\alpha, \qquad x_{\alpha_{i_1}} \widehat{\times \dots \times} x_{i_{\bar{k}_j}}^\gamma \in A^\star_{(\bar{j})} \subset A_{(\bar{j})}.$$

First of all, using the fact that the support of P_j is $A^\star_{(\bar{j})}$, let is establish the following statement.

ASSERTION. *Let B be an arbitrary subset in $Y_{(j)}$. For all $f \in C(Y)$ we have*

$$\|P_j f\|_{\tilde{H}_j \times B} \le \|P_j\| \|f\|_{A^\star_{(\bar{j}) \times B}} \le \|P_j\| \|f\|_{A(\bar{j}) \times B}. \tag{1.87}$$

First, let $f \in C(\widetilde{Y}_{(j)})$. In this case, let us establish the inequality

$$\|P_j f\|_{\widetilde{Y}_{(j)}} = \sup_{\tilde{y}_j \in \widetilde{Y}_{(j)}} |P_j f| \le \|P_j\| \|f\|_{A^\star_{(\bar{j})}}. \tag{1.88}$$

Surround points x_j^δ, $\delta = 1, \dots, N_j$ in the set $A^\star_{(j)}$ by disjoint neighborhoods V_j^δ. Define a function $F \in C(\widetilde{Y}_{(j)})$ that coincides with f on $A^\star_{(\bar{j})}$, vanishes on $\tilde{H}_{(j)} \setminus \bigcup_\delta V_j^\delta$ and is such that $\|F\|_{\widetilde{Y}_j} = \|f\|_{A^\star_{(\bar{j})}}$. Then, $P_j f = P_j F$ and

$$\|P_j f\|_{\widetilde{Y}_j} = \|P_j F\|_{\widetilde{Y}_j} \le \|P_j\| \|F\| = \|P_j\| \|f\|_{A^\star_{(\bar{j})}}.$$

Now, let $f(y) = f(y_j, \tilde{y}_j) \in C(Y)$. We have

$$\begin{aligned}
\|P_j f\|_{\widetilde{Y}_j \times B} &= \sup_{\substack{\tilde{y}_j \in \widetilde{Y}_j \\ y_j \in B}} \left| P_j f \left(\tilde{y}_j, \underbrace{y_j} \right) \right| = \sup_{y_j \in B} \left\| P_j f \left(\tilde{y}_j, \underbrace{y_j} \right) \right\| \\
&\le \sup_{y_j \in B} \|P_j\| \left\| f \left(\tilde{y}_j, \underbrace{y_j} \right) \right\|_{A^\star_{(\bar{j})}} = \|P_j\| \|f\|_{A^\star_{(\bar{j}) \times B}}.
\end{aligned}$$

The assertion is proved.

In particular, for $B = A_{(j)}$ as in (1.87) we obtain

$$\|P_j f\|_{\widetilde{Y}_j \times A_{(j)}} \le \|P_j\| \, \|f\|_A \qquad \left(A = A_{(\bar{j})} \times A_{(j)} \right) \tag{1.89}$$

and, moreover,

$$\|P_j f\|_A \le \|P_j\| \, \|A\|. \tag{1.90}$$

Now, let P be a projection from $C(Y)$ onto D obtained from P_j according to the formula (1.43). For all $w \in D$ in view of the formulas (1.71)–(1.72) we obtain

$$\begin{aligned}
& w = w_1 + \dots + w_m, \quad \text{where} \\
& w_1 = P_1 w, \\
& w_2 = P_2 (w - P_1 w) = P_2 (w - w_1), \\
& w_3 = P_3 (w - w_1 - w_2), \dots, \\
& w_j = P_j (w - w_1 - \dots - w_{j-1}), \dots, \qquad j = 1, \dots, m.
\end{aligned} \tag{1.91}$$

From (1.90) and (1.91) we obtain that

$$\|w_1\|_A \le \|P_1\| \, \|w\|_A, \quad \|w_2\|_A \le \|P_2\| \left(\|w\|_A + \|w_1\|_A \right) \le \|P_2\| \left(1 + \|P_1\| \right) \|w\|.$$

Continuing this chain, we arrive at a constant M_1 such that

$$\|w_j\|_A \leq M_1 \|w\|_A. \tag{1.92}$$

According to Lemma 1.17 (where we have to take $R = A_{(j)}$), for all j there exists a function $w_0' \in D_j$ such that

$$w_j' \Big|_{A_{(j)} \times \tilde{Y}_j} = w_j \Big|_{A_{(j)} \times \tilde{H}_j}, \tag{1.93}$$

$$\|w_j\| \leq \|w_j'\|_{A_{(j)} \times \tilde{Y}_j}. \tag{1.94}$$

Consider the function

$$w' = w_1' + \cdots + w_m'. \tag{1.95}$$

In any case, all equalities (1.93) hold in the set $A = A_{(j)} \times A_{(\tilde{j})}$; hence

$$w' \,|_A = w|_A\,.$$

In view of equicontinuity of the basis functions

$$\{h_j^\delta,\ j = 1, \ldots, m;\ \delta = 1, \ldots, N_j\}$$

and properness of the basis, we can assume Θ to be so large (and consequently all partitions of X_i, $i = 1, \ldots, N$, and $Y_{(\tilde{j})}$, $j = 1, \ldots, m$, so fine) that

$$\|w_j\|_{A_{(j)} \times \tilde{Y}_{(j)}} \leq 2\|w_j\|_{A_{(j)} \times A_{(\tilde{j})}} = 2\|w_j\|_A. \tag{1.96}$$

Combining (1.92)–(1.96), we obtain (1.85) with an appropriate constant M.

REMARK. The role played by points $x_i^1, \ldots, x_i^{n_i}$ that we are bound to include into A_i is not as significant as it may appear from the proof. Indeed, by considering the determinant (1.78) it is not hard to see that if we define the operators P_j not based on those points but on sufficiently close points $x_i'^1, \ldots, x_i'^{n_i}$, then the norms of the corresponding operators P_j' can be assumed to be bounded by constants that do not depend on the choice of points. And only this point is crucial for the proof. Therefore, if a partition Θ_i of the set X_i is sufficiently fine, then among the chosen points there will necessarily be points sufficiently close to $x_i^1, \ldots, x_i^{n_i}$ and whence we can choose points arbitrarily in each set of the partition Θ_i.

Totally separated variables. Approximation of measures in $D^\perp$ by finitely-supported measures. In the case of totally separated variables, Proposition 1.7 can be significantly refined.

THEOREM 1.19. *If in the structure of $D = D(Y)$ the variables are totally separated, then every measure $\mu \in D^\perp$, $\|\mu\| = 1$, is a weak $(\star)$ limit of a net of measures $\{\lambda_\theta\} \subset D^\perp$, each of which has a finite support and, moreover, $\|\lambda_\theta\| = 1$.*

For the proof we shall need the following lemma related to the duality of extremums.

LEMMA 1.20 ([**93**]). *Let G be a subspace in a Banach space B, let $V \subset B^\star$ be the linear hull of some linear functionals Φ_k, $\|\Phi_k\| = 1$, $k = 1, \ldots, t$, and $\Phi \subset V$. Then, the distance* $\operatorname{dist}(\Phi, V \cap G^\perp)$ *from Φ to $V \cap G^\perp$ satisfies the inequality*

$$\operatorname{dist}(\Phi, V \cap G^\perp) \leq \sup_{g \in G, |\Phi_k(g)| \leq 1} |\Phi(g)|. \tag{1.97}$$

PROOF. Consider a mapping $\Psi : B \to R^t$ defined by

$$g \in B \to \Psi(g) = (\Phi_1(g), \dots, \Phi_t(g)) \tag{1.98}$$

and let $G' = \Psi(G)$. In $(\mathbb{R}^t)^\star$ consider linear functionals $u = (u_1, \dots, u_t)$ on $\mathbb{R}^t$ that form $G'^\perp$. Clearly, $u \in G'^\perp$ if and only if

$$u_1\Phi_1(g) + \dots + u_t\Phi_t(g) = 0. \tag{1.99}$$

Thus, $u \in G'^\perp \Leftrightarrow \sum_1^t u_k\Phi_k \in G^\perp \cap V$. In $\mathbb{R}^t$, take the ℓ^∞-norm. Then, $(\mathbb{R}^t)^\star$ is equipped with the ℓ^1-norm. Let $\Phi = \sum_1^m a_k\Phi_k$. To the functional Φ there corresponds in $(\mathbb{R}^t)^\star$ the functional $a = (a_1, \dots, a_t)$. Applying the duality relation, we obtain

$$\begin{aligned}\operatorname{dist}\left(\Phi, V \cap G^\perp\right) &= \inf_{\varphi \in V \cap G^\perp} \|\Phi - \varphi\| = \inf_{u \in G'^\perp} \left\|\sum_1^t (a_k - u_k)\,\Phi_k\right\| \\ &\le \inf_{u \in G'^\perp} \sum_1^t |a_k - u_k| - \operatorname*{dist}_{(\ell_1)}\left(a, G'^\perp\right) \\ &= \sup_{\substack{g' \in G' \\ \|g'\|_\infty \le 1}} |a(g')| = \sup_{g \in G, |\Phi_k(g)| \le 1, k=1,\dots,t} |\Phi(g)|.\end{aligned}$$

PROOF OF THEOREM 1.19. According to Proposition 1.7, for $\mu \in D^\perp$, $\|\mu\| = 1$, there exists a net of measures $\{\mu_\theta\}$, $\|\mu_\theta\| = 1$, weak $(\star)$ converging to μ. The order of indices is determined by the set of partitions $\{\theta\}$ of the basis sets X_i. The measure μ_θ is supported on a set that coincides with the set $A = A^\theta$ in Theorem 1.18. Since, in view of Theorem 1.12, the system of basis functions is proper, Proposition 1.7 shows that for $\theta > \theta_0(\varepsilon)$ we have

$$\|\mu_\theta\|_D < \varepsilon. \tag{1.100}$$

Estimate the distance d in $C(Y)^\star$ from μ_θ to the subspace of measures supported on A^θ and annihilating the subspace D. In order to use Lemma 1.20 we set $B = C(Y)$, $G = D$, $\Phi_k = \widehat{x}_k$, $x_k \in A^\theta$ (t is the number of points in A^θ), $\Phi = \mu_\theta$. Then, by Lemma 1.20 we have

$$d \le \sup_{\substack{w \in D \\ |w|_{A^\theta} \le 1}} |\mu_\theta(w)| = \sup_{\substack{w \in D \\ \|w\|_{A^\theta} \le 1}} |\mu_\theta(w)|. \tag{1.101}$$

According to Theorem 1.18, for $w \in D$ there exists $w' \in D$ such that

$$w\,|_{A^\theta} = w'|_{A^\theta}, \qquad \|w'\| \le M\|w\|_{A^\theta}. \tag{1.102}$$

Therefore, if $\|w\|_{A^\theta} \le 1$, then $\|w'\| \le M$ and in view of (1.101), (1.102) we have

$$d \le \sup_{\substack{w \in D \\ \|w\|_{A^\theta} \le 1}} |\mu_\theta(w)| = \sup_{\substack{w' \in D \\ \|w'\| \le M}} |\mu_\theta(w')| \le M\varepsilon. \tag{1.103}$$

Thus, there exists a measure $\lambda_\theta \in D^\perp$ supported on A^θ and such that

$$\|\mu_\theta - \lambda_\theta\| \leq M\varepsilon.$$

It is clear that the net $\left\{ \frac{\lambda_\theta}{\|\lambda_\theta\|} \right\}$ satisfies all the requirements.

Subspaces of sums. Let all $H(\widetilde{Y}_j)$ be one-dimensional and consist of constants, $j = 1, \dots, m$. Then, the subspace D consists of functions

$$\sum_{j=1}^{m} g_j\left(y_{(j)}\right), \qquad g_j \in C\left(Y_j\right). \tag{1.104}$$

As above, assume that we cannot have inclusions $(j) \subset (k)$ for $j \neq k$, although it is possible to have $(j) \cap (k) \neq \emptyset$. A system biorthogonal to the basis of $H(\widetilde{Y}_j)$ consists of any measure μ_j on $\widetilde{Y}_j$ for which

$$\int_{\widetilde{Y}_j} d\mu_j = 1. \tag{1.105}$$

Projections $P_j : C(Y) \to D_j = \{g_j\left(y_j\right)\}$ and the complimentary projections R_j are given by the formulas:

$$P_j : f\left(y_{(j)}, y_{(\tilde{j})}\right) \to \int_{\widetilde{Y}_j} f\left(\underbrace{y_{(j)}}, y_{(\tilde{j})}\right) d\mu_j; \qquad R_j : f(y) \to f(y) - P_j f. \tag{1.106}$$

In particular, for μ_j we can take the δ-measure at a fixed point $y^\circ_{(\tilde{j})} \in \widetilde{Y}_{(j)}$, and then

$$P_j : f(y) \to f\left(y_j, y^\circ_{(\tilde{j})}\right), \qquad R_j : f(y) \to f(y) - f\left(y_{(j)}, y^\circ_{(\tilde{j})}\right) \tag{1.107}$$

are extensions of the formulas (1.48)–(1.49). Since the assumption of total separation of variables is obviously satisfied when D has form (1.104), we can formulate the following corollary.

COROLLARY 1.21. *The operator P constructed out of operators P_j ((1.106) or 1.(107)) according to* (1.43) *is a projection of $C(Y)$ onto D that consists of functions* (1.104). *The subspace D consists of those and only those functions $F \in C(Y)$ for which* (*see* (1.45))

$$RF = 0, \quad \text{where } R = R_1 \dots R_m, \; R_j = I - P_j, \; j = 1, \dots, m. \tag{1.108}$$

For all $f \in C(Y)$ and $F = Pf$ the interpolational identities hold:

$$\int_{\widetilde{Y}_j} \left[f\left(\underbrace{\left(y_{(j)}\right.}, y_{(\tilde{j})}\right) - F\left(y_{(j)}, y_{(\tilde{j})}\right)\right) \right] d\mu_j = 0, \qquad j = 1, \dots, m, \tag{1.109}$$

and, in particular, for the case (1.107),

$$f\left(y_j, \tilde{y}^\circ_j\right) - F\left(y_j, y^\circ_j\right) = 0. \tag{1.110}$$

Bounded functions. Now let us deal with bounded functions instead of continuous ones. $B(T)\,(\ell^\infty(T))$ is the space of bounded functions with the uniform norm on a set T. Let X, X_i, $i = 1, \dots, N$, be arbitrary sets, $\Phi_i : X \to X_i$ some mappings. By BD we denote the subspace in $B(X)$ that consists of functions

$$\sum_{i=1}^{N} h^i(x) g_i \circ \Phi_i(x), \tag{1.111}$$

with $g_i \in B(X_i)$ and the basis functions $h^i(x) \in B(X)$. We obtain an analogue of the subspace in (1.1). Similarly, we form analogues of the subspace D in the situations (1.4), (1.7), (1.11)–(1.12), (1.51), and (1.68):

The analogue of (1.4):

$$\sum_{1}^{N} h^i(x) g_i(x_i), \qquad h^i \in B(Y), \quad g_i \in B(X_i). \tag{1.112}$$

The analogue of (1.7):

$$\sum_{j=1}^{m} h^j(x) g_j(y_j), \qquad h^j \in B(Y), \quad g_j \in B(Y_j). \tag{1.113}$$

The analogue of (1.11), (1.68):

$$BD = \sum_{j=1}^{m} H^j \otimes B(Y_j), \qquad H^j = H^j\left(\widetilde{Y}_j\right) \subset B\left(\widetilde{Y}_j\right). \tag{1.114}$$

Here, as in (1.11), the H^j are finite dimensional subspaces.

The analogue of (1.12):

$$\sum_{j=1}^{m} \sum_{\delta=1}^{m_j} h_j^\delta(\tilde{y}_j)\, g_j^\delta(y_j), \qquad h_j^\delta \in B\left(\widetilde{Y}_j\right), \quad g_j^\delta \in B(Y_j). \tag{1.115}$$

The analogue of (1.51):

$$BD_j = H^j \otimes B(Y_j), \qquad H^j \subset B\left(\widetilde{Y}_j\right). \tag{1.116}$$

(We have given a complete list of those subspaces for convenience in future references.) The definition of a proper basis system $\{h^j\}$ does not change, nor do those of separated and totally separated variables.

PROPOSITION 1.22. *The statements of Proposition* 1.1, *Lemma* 1.10, *Theorem* 1.12, *and Corollary* 1.2 *also hold for the subspace* BD.

Since the space $B(T)$ is dual to $\ell^1(T)$ (see §2 in Chapter 1), Lemmas 1.2–1.5, Propositions 1.6–1.7, and Theorem 1.19 can be applied to the description of the annihilator $(BD)^\perp$ that consists of all $\mu \in \ell^1(T)$ for which $\langle \mu, w\rangle = 0$, $w \in BD$ ($\langle \mu, w\rangle$ denotes the action of the functional w on the element μ, i.e., the integral of the function w with respect to the measure μ). Weak ($\star$) convergence in Proposition 1.7 and Theorem 1.19 means convergence over every continuous function, which is weaker than weak convergence in the space $\ell^1(T)$ (that is, convergence over all bounded functions).

Bibliographical notes. Subspaces like those in (1.12) in connection with approximation of functions of several variables (with various goals) have been considered in many papers. In particular, for H^j spaces of ordinary polynomials of a given degree and splines were used.

Consideration of the projections P_j in (1.53), (1.69), and the projection P constructed from P_j using (1.43) is the essence of the "blending" method developed in a number of papers by Gordon; see, e.g., [**66**], [**67**]. In these papers formulas (1.43), (1.45), (1.71), (1.72) are widely used together with other relations associated with the Boolean distributive structure of commuting (and in some cases even noncommuting—e.g., [**68**]) projections. Blending (of variables) occurs when a projection $P_j : C(X_j) \to H^j(X_j)$ originally defined in a space of functions of one variable is extended to the whole space $C(Y)$ of functions of several variables, and the variables different from x_j are treated as parameters.

For applications to numerical methods in boundary value problems for differential equations, operators similar to P_j and P are considered in spaces of differentiable functions. For D_j one can take the subspace of solutions of the ordinary differential equation $d_j f = 0$ with the linear differential operator d_j (with respect to the variable x_j); and then, according to (1.45), D consists of the solutions of the partial differential equations $d_1 \dots d_m f = 0$.

Linear functionals used in constructing the projections P_j are chosen so that interpolation conditions (1.54), (1.70) would be helpful for a boundary value problem. The method is also used for numerical integration and solving integral equations, and some other problems.

Brudnyi [**24**] arrived at the considerations associated with the Boolean structure of the set of projections almost at the same time; he was investigating subspaces similar to (1.12) in connection with approximation of various classes of functions given by their differential properties. The approximation was conducted not only in the uniform metric, but in the integral metrics as well. For basis functions h_j^δ in [**24**] ordinary algebraic or trigonometric polynomials were taken, and every product $Y_{(j)}$ contained the same number of factors X_i that provided a "homogeneity" of functions in D. For the latter situation, problems of uniform approximation by the subspace D were also studied by Vaindiner [**139**]. One of the main results in [**139**] was establishing Corollary 1.15 for that case, yet a proof was not given there. Theorems 1.18 and 1.19 are most important from our viewpoint. For the case of two variables they have been proved by Cheney and Light [**93**], together with Lemmas 1.17 and 1.20, and Proposition 1.6.

The presentation in this section mainly followed the scheme in [**93**], although the general case naturally turned out to be more complicated.

§2. On the existence of best approximations

THEOREM 2.1. *If a basis system is proper in $B(Q)$, then the subspaces in* (1.111)–(1.116) *are closed in the weak* ($\star$) *topology of the space $B(Q)$, $Q \subset Y$* ($B(X)$ *for* (1.111)), *and are proximinal. In particular, the subspace BD* (1.114) *is weak* ($\star$) *closed and proximinal in $B(Y)$, provided that the variables are totally separated.*

PROOF. Since the subspace $B(Q)$ is dual to $\ell^1(Q)$, we need only prove weak ($\star$) closedness. Keep the notation of (1.113). Let $\{w_\theta\}$ be a net of functions in $BD(Q)$, i.e., functions

$$w_\theta = \sum_{j=1}^{m} h^j(x) g_{j,\theta}(y_j), \qquad g_{j,\theta} \in B(Y_j),$$

that converges in the weak $(\star)$ topology to a function w, and $\|w_\theta\| \le M$. Since the $\|x_\theta\|$ are bounded and the basis h^j is a proper system, the $\|g_{j,\theta}\|$, $j = 1, \dots, m$, are also bounded. In view of the Alaoglu theorem, taking a subnet if necessary, assume that $\{g_{j,\theta}\}$ converges in the weak $(\star)$ topology of $B(Y_j)$ to $g_j \in B(Y_j)$, $j = 1, \dots, m$. But then it is clear that

$$w = \sum_{1}^{m} h^j(x) g_j(y_j) \in BD, \qquad \|w\| \le M.$$

We have proved weak $(\star)$ closedness of the intersection of $BD(Q)$ with any closed ball. But then $BD(Q)$ is weak $(\star)$ closed (see, e.g., [**40**]), and the proof is complete. According to Proposition 1.22 and Theorem 1.12, the assumption (1.114) about total separation of variables yields properness of the basis system in $B(Y)$.

COROLLARY 2.2. *The subspace of sums*

$$\sum_{j=1}^{m} g_j\left(y_{(j)}\right), \qquad g_j \in B(Y_j), \quad (j) \not\subset (k) \text{ provided that } j \neq k \tag{2.1}$$

is closed in the weak $(\star)$ *topology of* $B(Y)$ *and is proximinal* (*cf.* §3 *in Chapter* 2).

Approximation by sums on sets $Q \subset Y$. Let $Q \subset Y = X_1 \times \cdots \times X_N$ be a set. Naturally, properties of the subspace $BD(Q)$ of restrictions to Q of functions from $BD(Y)$ depend on geometry of Q. We consider these questions for a subspace made out of functions (2.1), thus continuing certain considerations in §§2, 3 in Chapter 2.

Bars (cf. §§ 2.3 in Chapter 2). Let (j) be a set of indices in $(1, \dots, N)$ and $y^\circ_{(j)} \in Y_{(j)}$, a fixed point. By $Q_{y^\circ_{(j)}}$ we denote the cross-section of Q by the hyperplane $y_{(j)} = y^\circ_{(j)}$. Thus,

$$Q_{y^\circ_{(j)}} = \{(y^\circ_{(j)}, y_{(\tilde{j})}) \in Q\}. \tag{2.2}$$

(Of course, for a given point $y^\circ_{(j)} \in Y_{(j)}$, $Q_{y^\circ_{(j)}}$ may turn out to be empty.) Let (k) be another set of indices and $(k) \subset (\tilde{j})$. (Recall that $(\tilde{j})$ is the complementary set to (j).) By a bar of the passport $[(j),(k)]$, or simply a $[(j),(k)]$-bar, we mean a cross-section $Q_{y^\circ_{(j)}}$ such that

$$\pi_{(k)}\left[Q_{y^\circ_{(j)}}\right] = \pi_{(k)}[Q]. \tag{2.3}$$

Of course, such bars do not necessarily exist. The following lemma is obvious.

LEMMA 2.3. *If a cross-section $Q_{y^\circ_{(j)}}$ is a bar of the passport $[(j),(k)]$, then it is also a bar of a passport $[(j),(r)]$ whenever $(r) \subset (k)$.*

PROOF. According to the assumptions, (2.3) holds. But since $(r) \subset (k)$, we have

$$\pi_{(r)} Q_{y^\circ_{(j)}} = \pi_{(r)} \circ \pi_{(k)} Q_{y^\circ_{(j)}} = \pi_{(r)} Q.$$

THE MAIN LEMMA. Let (j) and (k) be two sets of indices. We shall say that there exists a bar of the passport $[(j),(k)]^\star$ in Q if Q contains at least one bar of either the passport $[(j)\backslash(k),(k)]$ or the passport $[(k)\backslash(j),(j)]$

LEMMA 2.4. *Let Q contain a bar of the passport $[(j),(k)]^\star$. Then the following hold:*

1. *If a net of functions $\{F_\alpha\}$,*

$$F_\alpha = \varphi_\alpha(y_{(j)}) + \psi_\alpha(y_{(k)}), \tag{2.4}$$

converges to a function F in the weak $(\star)$ topology of the space $B(Q)$, then φ_α and ψ_α can be replaced by functions $\varphi'_\alpha\left(y_{(j)}\right)$ and $\psi'_\alpha\left(y_{(k)}\right)$ such that

$$F_\alpha = \varphi'_\alpha\left(y_{(j)}\right) + \psi'_\alpha\left(y_{(k)}\right) \tag{2.5}$$

and, moreover, $\{\varphi'_\alpha\left(y_{(j)}\right)\}$ and $\{\psi'_\alpha\left(y_{(k)}\right)\}$ weak $(\star)$ converge.

2. *A subspace of functions*

$$F = \varphi\left(y_{(j)}\right) + \psi\left(y_{(k)}\right) \tag{2.6}$$

is closed in the weak $(\star)$ topology of $B(Q)$, and therefore is proximinal.

PROOF. Let (r) be a set of indices. First of all, it is clear that if a net of functions $\{\Phi_\alpha\left(y_{(r)}\right)\}$ weak $(\star)$ converges in $B(Q)$ to a function Φ, then Φ depends on $y_{(r)}$ only. Indeed, let $(y_{(r)}, y^1_{(N)\backslash(r)})$ and $(y_{(r)}, y^2_{(N)\backslash(r)})$ be two points that differ only by coordinates with indices inside $(N)\backslash(r)$. Then

$$\Phi\left(y_{(r)}, y^1_{(N)\backslash(r)}\right) = \lim_\alpha \Phi_\alpha\left(y_{(r)}\right) = \Phi\left(y_{(r)}, y^2_{(N)\backslash(r)}\right),$$

so that Φ depends only on $y_{(r)}$. Now, let $\{F_\alpha\}$ be a net of the form mentioned in the lemma. Consider the functions

$$\begin{aligned} \varphi'_\alpha\left(y_{(j)}, y_{(k)\backslash(j)}\right) &= F_\alpha\left(y_{(j)}, y_{(k)\backslash(j)}\right) - \psi'_\alpha\left(y_{(k)}\right), \\ \psi'_\alpha\left(y_k\right) &= F\left(y^\circ_{(j)\backslash(k)}, y_{(j)\cap(k)}, y_{(k)\backslash(j)}\right). \end{aligned} \tag{2.7}$$

Here, $y^\circ_{(j)\backslash(k)}$ is the point for which the cross-section $Q_{y^\circ_{(j)\backslash(k)}}$ is a bar of the passport $[(j)\backslash(k),(k)]$. (We conduct the argument for that case. When there exists a bar of the passport $[(k)\backslash(j),(j)]$, the arguments follow modulo obvious symmetric changes with respect to (k) and (j).) From (2.7) it is clear that $\psi'_\alpha\left(y_{(k)}\right)$ indeed depends only on the variable $y_{(k)}$. Let us show that the function φ'_α in (2.7) depends only on $y_{(j)}$ and not on $y_{(k)\backslash(j)}$. Take two points in Q where

$$\begin{aligned} t_1 &= \left(y_{(j)}, y^1_{(k)\backslash(j)}, y^1_{(s)}\right) = \left(y_{(j)\backslash(k)}, y_{(j)\cap(k)}, y^1_{(k)\backslash j}, y^1_{(s)}\right), \\ t_2 &= \left(y_{(j)}, y^2_{(k)\backslash(j)}, y^2_{(s)}\right) = \left(y_{(j)\backslash(k)}, y_{(j)\cap(k)}, y^2_{(k)\backslash(j)}, y^2_{(s)}\right), \\ &\text{where } (s) = \widetilde{(j)\cup(k)} \quad \text{is the complement to } (j)\cup(k). \end{aligned} \tag{2.8}$$

These points differ by coordinates with indices in $(k)\backslash(j)$ and (s), but the latter are unimportant in view of the above remark. Since $Q_{y^\circ_{(j)\backslash(k)}}$ is a bar of the passport

$[(j)\backslash(k),(k)]$, there exist two points

$$
\begin{aligned}
t_3 &= \left(y^{\circ}_{(j)\backslash(k)}, y_{(j)\cap(k)}, y^{1}_{(k)\backslash(j)}, y^{3}_{(s)}\right), \\
t_4 &= \left(y^{\circ}_{(j)\backslash(k)}, y_{(j)\cap(k)}, y^{2}_{(k)\backslash(j)}, y^{4}_{(s)}\right),
\end{aligned} \tag{2.9}
$$

whose coordinates with indices in (k) coincide with those of t_1 and t_2, respectively. For a function Φ on Q set

$$\Delta\Phi = \Phi(t_1) - \Phi(t_2) + \Phi(t_4) - \Phi(t_3). \tag{2.10}$$

Since ψ_α and ψ'_α only depend on $y_{(k)}$, whereas φ_α depends only on $y_{(j)}$, we find at once that

$$\Delta\psi_\alpha = 0, \qquad \Delta\psi'_\alpha = 0, \qquad \Delta\varphi_\alpha = 0, \qquad \Delta F_\alpha = 0.$$

Therefore, $\Delta\varphi'_\alpha = \Delta F_\alpha - \Delta\varphi'_\alpha = 0$ as well. But the latter equality can be rewritten in the form

$$\varphi'_\alpha\left(y_{(j)}, y^{1}_{(k)\backslash(j)}\right) = \varphi'_\alpha\left(y_{(j)}, y^{2}_{(k)\backslash(j)}\right). \tag{2.11}$$

Thus, φ'_α is a function of $y_{(j)}$ only. Since $\{F_\alpha\}$ converges weak $(\star)$, so does $\{\psi'_\alpha\}$. But then $\{\varphi'_\alpha\}$ also converges in this topology. Thus, using once again the remark made in the beginning of the argument, we have proved that

$$F = \lim_\alpha F_\alpha = \lim_\alpha \varphi'_\alpha + \lim_\alpha \psi'_\alpha = \varphi\left(y_{(j)}\right) + \psi\left(y_{(k)}\right).$$

This completes the proof of the lemma.

Depending on the structure of the sums (2.1), one can by Lemma 2.4 choose various geometric requirements on Q that would provide weak $(\star)$ closedness of the subspace $BD(Q)$ and hence its proximinality. Let us give one example.

Geometry of Q and proximinality of a subspace of sums.

THEOREM 2.5 ([**83**]). *Let Q contain bars of the following passports:*

$$
\begin{array}{lr}
[(2)\cup(3)\cup\cdots\cup(m),(1)]^\star : & (2.12_1) \\
[(3)\cup(4)\cup\cdots\cup(m),(2)]^\star : & (2.12_2) \\
\cdots\cdots\cdots\cdots\cdots\cdots\cdots\cdots & \\
\qquad [m,m-1]^\star & (2.12_{m-1})
\end{array} \tag{2.12}
$$

Then the subspaces of functions (2.1) *is weak* $(\star)$ *closed and proximinal in* $B(Q)$.

PROOF. Let a net of functions $\{F_\alpha\}$,

$$F_\alpha = \varphi^1_\alpha\left(y_{(1)}\right) + \varphi^{(2)}_\alpha\left(y_{(2)}\right) + \cdots + \varphi^m_\alpha\left(y_{(m)}\right), \tag{2.13}$$

converge weak $(\star)$ in $B(Q)$. In view of (2.12_1) and the preceding lemma we have

$$F_\alpha = \varphi^{1'}_\alpha + F^1_\alpha,$$

where, in view of the structure of F_α (formula (2.13)), F^1_α has the form

$$F^1_\alpha = {\varphi'_\alpha}^2\left(y_{(2)}\right) + \cdots + {\varphi'_\alpha}^m\left(y_{(m)}\right), \tag{2.14}$$

while $\{\varphi'^{1}_{\alpha}(y_{(1)})\}$ and $\{F^1_\alpha\}$ converge in the weak $(\star)$ topology. Because of the existence of the bar (2.12_2), we again apply Lemma 2.4, etc. After the $(m-1)$st step we obtain that

$$F_\alpha = \psi^1_\alpha\left(y_{(1)}\right) + \psi^2_\alpha\left(y_{(2)}\right) + \cdots + \psi^m_\alpha\left(y_{(m)}\right)$$

$(\psi^1_\alpha = \varphi'^{1}_{\alpha}$, etc.), and each one of the nets $\{\psi^k_\alpha\left(y_{(k)}\right)\}$, $k = 1, \dots, m$, converges weak $(\star)$.

The universal set of bars. Let us point out a sufficient condition which guarantees that for any set of index groups (j), $(j) = 1, \dots, m$, the subspace of sums (2.1) is proximinal.

THEOREM 2.6. *For all $i = 1, \dots, N$, let there exist x_i° such that the cross-section $Q_{x_i^\circ}$ of the set Q by the hyperplane $x_i = x_i^\circ$ is a bar of the passport $[i, (1, \dots, i-1, i+1, \dots, N)]$. Then, for any set of index groups (j), the subspace BD of functions* (2.1) *is closed in the weak $(\star)$ topology and is proximinal.*

A proof is obtained by using Lemma 2.3 in order to show that in Q there exist bars for all passports and then applying Theorem 2.5.

Approximination of continuous functions by sums. The question concerning existence of the best approximation in the space of continuous functions is more complicated.

THEOREM 2.7. *The subspace of sums* (1.104) *is proximinal in $C(Y)$, provided that any two sets of indices (j) and (k), $j \neq k$, are disjoint. For a function $f \in C(Y)$ there exists a best approximation in the subspace* (1.104) *that has modulus of continuity over each group of variables $y_{(j)}$ not worse than that of f.*

PROOF. Consider approximation of f in $B(Y)$ by a subspace (2.1) that differs from (1.104) in that it allows bounded, not merely continuous functions. According to Theorem 2.5, f has the best approximation

$$\varphi_1\left(y_{(1)}\right) + \cdots + \varphi_m\left(y_{(m)}\right) \in BD(Y).$$

Consider a family of functions of $y_{(1)}$:

$$F\left(y_{(1)}, \dots, y_{(m)}\right) = f\left(y_{(1)}, \underbrace{y_{(2)}, \dots, y_{(m)}}\right) - \sum_{j=2}^{m} \varphi_j\left(y_{(j)}\right)$$

for which the variables $y_{(2)}, \dots, y_{(m)}$ are parameters. All functions in the family are equicontinuous with respect to $y_{(1)}$. Set

(2.15)

$$\varphi_1^\star\left(y_{(1)}\right) = \frac{1}{2}\left[\sup_{y_{(2)}, \dots, y_{(m)}} F\left(y_{(1)}, \dots, y_{(m)}\right) + \inf_{y_{(2)}, \dots, y_{(m)}} F\left(y_{(1)}, \dots, y_{(m)}\right)\right].$$

Then, for all $y_{(1)}$ we have

$$\begin{aligned}&\sup_{y_{(2)} \cdots y_{(m)}} \left|F\left(y_{(1)}, \dots, y_{(m)}\right) - \varphi_1^\star\left(y_{(1)}\right)\right| \\ &\qquad \leq \sup_{y_{(2)} \cdots y_{(m)}} \left|F\left(y_{(1)}, \dots, y_{(m)}\right) - \varphi_1\left(y_{(1)}\right)\right|,\end{aligned}$$

and hence

$$\|f - \varphi_1^\star(y_{(1)}) - \varphi_2(y_2) - \cdots - \varphi_m(y_{(m)})\| = \sup_{y_{(1)}} \sup_{y_{(2)} \cdots y_{(m)}} |F - \varphi_1^\star(y_{(1)})|$$

$$\leq \sup_{y_{(1)}} \sup_{y_{(2)} \cdots (m)} |F - \varphi_1(y_{(1)})| = \left\| f - \sum_1^m \varphi_j \right\|.$$

Therefore, the sum $\varphi_1^\star(y_{(1)}) + \varphi_2(y_{(2)}) + \cdots + \varphi_m(y_{(m)})$ also gives the best approximation to f in $B(Y)$. Now, starting with it, we deal with the second group of variables $y_{(2)}$, etc.

We obtain the best approximation of f in $B(Y)$ by a function $\varphi_1^\star(y_{(1)}) + \cdots + \varphi_m^\star(y_{(m)})$ with continuous summands. Moreover, from (2.15) and similar formulas for other summands and properties of the operator M (see Lemma 6.13 in Chapter 2) follow the statements concerning the nature of continuity of the best approximation.

Theorem 2.7 was established in [**113**], and it generalizes the existence theorem of Diliberto and Straus, and Kolmogorov for two variables (Theorem 6.14 in Chapter 2).

Duality relations. Theorem 1.19 allows us to somewhat simplify ordinary duality relations for that situation.

PROPOSITION 2.8. *For the subspace D* (1.11) *with totally separated variables* (1.13) *and $f \in C(Y)$, we have*

$$\operatorname{dist}(f, D) = \sup_{\substack{\mu \in D^\perp \\ \|\mu\| \leq 1}} |\mu(f)| = \sup_{\substack{\mu \in D^\perp \cap \ell_k^1 \\ \|\mu\| = 1}} |\mu(f)|. \tag{2.16}$$

Here, $\operatorname{dist}(f, D)$ is the distance from f to D, and ℓ_k^1 (as in §3 in Chapter 1) is the set of measures with finite support.

Equality of distances from a continuous function to subspaces D and BD. In a more general situation than that in Theorem 1.19 one can successfully apply the not-so-sophisticated Proposition 1.7.

THEOREM 2.9. *Let $X_1, \ldots, X_N$ be compact sets, $h^1(x), \ldots, h^M(x)$ be continuous functions on Y that form a basis in D and BD, where D and BD are the subspaces defined by* (1.7) *and* (1.112), *respectively. For $f \in C(Y)$, we have*

$$\operatorname{dist}(f, D) = \operatorname{dist}(f, BD). \tag{2.17}$$

PROOF. According to the ordinary duality relation, there exists a regular Borel measure $\mu \in D^\perp$, $\|\mu\| = 1$, such that

$$\mu(f) = \operatorname{dist}(f, D). \tag{2.18}$$

According to Proposition 1.7 we can construct a net $\{\mu_\theta\}$ of measures with finite supports that converge weak $(\star)$ to μ in $C(Y)^\star$. Each measure μ_θ is in ℓ^1, hence acts on all bounded functions. It is easy to see that the relation (1.32) proven in Proposition 1.7, saying that $\lim_\theta |\mu_\theta|_D = 0$, can be replaced by a stronger one:

$$\lim_\theta |\mu_\theta|_{BD} = 0. \tag{2.19}$$

Indeed, in the proof of (1.32) we only used continuity of h^i and boundedness of g_j. Let $w_0 \in BD$ be the best approximation to f in BD. It exists because $BD(Y)$ is proximinal (Theorem 2.5). Since

$$\mu_\theta(f) = \mu_\theta(f - w_0) + \mu_\theta(w_0), \qquad |\mu_\theta(w_0)| \le |\mu_\theta|_{BD}|w_0| \to 0,$$

we obtain

$$\begin{aligned}\operatorname{dist}(f, D) = \mu(f) &= \lim_\theta \mu_\theta(f) = \lim_\theta \mu_\theta(f - w_0)\\ &\le \|\mu_\theta\|\|f - w_0\| = \|f - w_0\| = \operatorname{dist}(f, BD).\end{aligned}$$

Since the converse inequality is obvious, the theorem is proved.

REMARK. In [**92**], [**94**] the theorem was proved for $N = 2$ and in a less general situation (separated variables for $N = 2$ are always totally separated) by totally different arguments based on the Michael theorem on continuous selections (see, e.g., [**105–106**]). Attempts to apply these arguments to our situation would have required more severe restrictions than in Theorem 2.9: it is necessary that any two groups of indices (j) and (k) be disjoint when $j \neq k$.

Further results on existence of best approximations. Convenient results based on properties of a function that would guarantee existence of the best approximation in a general subspace (1.1), (1.7), or (1.11)–(1.14) are still lacking in the case of $N > 2$ variables, even for the simplest geometric case when approximation is carried out on a product $X_1 \times \cdots \times X_N$. For $N = 1$ (subspaces $H(X_1) \otimes C(X_2)$ in $C(X_1 \times X_2)$) and $N = 2$, a survey of results can be found in Chapter 2 of [**94**], where metrics different from C and ℓ^∞ are also considered. Among the investigations that have appeared since that book was published, let us point out the following negative result.

THEOREM 2.8 ([**46**]). *The subspace*

$$D = H^1(X_1) \otimes C(X_2) + H^2(X_2) \otimes C(X_1),$$

where $X_1 = X_2 = [0, 1]$, H^1, H^2 *are two-dimensional spaces of linear polynomials, is not proximinal in* $C(X_1 \times X_2)$.

Also, let us point out a survey [**50**] where the connection of those problems to the theory of Chebyshev centers is given.

In the case when D and BD consist of sums (all basis functions $h^j \equiv 1$), in [**54a**] geometric conditions for sets $Q \subset X_1 \times \cdots \times X_N$ that guarantee proximinality of D and BD in $C(Q)$ and $B(Q)$, respectively, were studied. These conditions generalize criteria from §3 of Chapter 2, where the results of [**54**] were presented. The geometric conditions considered in [**54a**] are of a different character than those used in Theorem 2.5.

In [**88**] Konyagin gave a rather general result concerning proximinality in L^p, $1 \le p \le \infty$ (in particular, in ℓ^∞), of sets of a certain class defined by special functional equations. Some linear subspaces, linear superpositions, and also some nonlinear sets fall into that class. The spaces L^p themselves are products and, in particular, ℓ^∞ is considered on $X_1 \times \cdots \times X_N$.

§3. Effective construction of best approximations

On Chebyshev's theorem. The complexity of the geometry of the supports of measures in $D^\perp$ is the reason why direct analogues of Theorem 6.1 in Chapter 2 are lacking for the approximation by the subspaces (1.7)–(1.13), or even by subspaces of sums (1.104) when the number of variables $N > 2$. Of course, general methods of characterization of the best approximation via functionals in $D^\perp$ apply (see Theorem 6.2 in Chapter 2); however, their specialization that would take into account the structure of the subspace D is hampered by the geometry of the supports of measures in $D^\perp$. Let us give an ε-version of a characterization of the best approximation that may be useful in view of Lemmas 1.2–1.4.

PROPOSITION 3.1. *Let D satisfy the assumptions of Theorem 1.19. In order that a function $\varphi \in D$ give the best approximation to a given function $f \in C(Y)$ it is necessary and sufficient that for each $\varepsilon > 0$ there exists a measure $\mu_\varepsilon \in D^\perp$, $\|\mu_\varepsilon\| = 1$, that has a finite support S_{μ_ε} and satisfies the condition*

$$|\mu_\varepsilon(f)| \geq \|f - \varphi\| - \varepsilon. \tag{3.1}$$

PROOF. If $\varphi \in D$ is the best approximation to f, then according to Theorem 6.2 in Chapter 2 there exists a measure $\mu \in D^\perp$, $\|\mu\| = 1$, with $\mu(f) = \|f - \varphi\|$. For μ_ε it suffices to take a measure with an appropriate index from the sequence $\{\mu_\theta\}$ described by Theorem 1.1 weak $(\star)$ converging to μ. If for μ_ε (3.1) holds and $\|\mu_\varepsilon\| = 1$, then for all $\varphi \in D$ we have

$$|\mu_\varepsilon(f)| = |\mu_\varepsilon(f - \psi)| \leq \|f - \psi\|, \tag{3.2}$$

and combining (3.2) and (3.1) yields that $\|f - \varphi\| \leq \|f - \psi\|$.

In the general theory of best approximation there is another type of characterization theorem different from Theorem 6.2 in Chapter 2, the Kolmogorov criterion (see, e.g., [**120**]). Naturally, it applies to our problems as well. However, its specialization that would use effectively the specific structure of the subspace D is still difficult.

For certain classes of functions in case of a subspace of sums one can directly point out best approximations. We now present an interesting result of Babaev ([**19**], [**21**]).

Monotonicity of functions of several variables. Let Y be an m-dimensional parallelepiped in $\mathbb{R}^m$:

$$Y = \{x \in \mathbb{R}^m : a_i \leq x \leq b_i,\ i = 1, \dots, m\}, \tag{3.3}$$

which we shall denote by $\Pi(a, b)$ or, in a more expanded form,

$$\begin{bmatrix} b_1 \dots b_m \\ a_1 \dots a_m \end{bmatrix}.$$

Let $P_1, \dots, P_{2^m}$ be the vertices of $\Pi(a, b)$. For functions $f \in C(\Pi(a, b))$ consider the linear functional

$$L(f, \Pi(a, b)) = 2^{-m} \sum_{k=1}^{m} (-1)^{\delta(P_k)} f(P_k), \tag{3.4}$$

where $\delta(P_k)$ is the number of coordinates a_i for the vertex P_k. (For example, $\delta(b_1, \dots, b_m) = 0$, $\delta(a_1, \dots, a_m) = m$, $\delta(a_1, a_2, a_3, b_4, \dots, b_m) = 3$.) Clearly, $\|L\| = 1$.

For an arbitrary parallelpiped $\Pi(\alpha, \beta) \subset \Pi(a, b)$ we can similarly construct a functional $L(f, \Pi(\alpha, \beta))$. If we set $h_i = b_i - a_i$, $i = 1, \dots, m$, it is easy to calculate that

$$L(f, \Pi(a, b)) = 2^{-m} \Delta_{h_1 h_2 \dots h_m}(f), \tag{3.5}$$

where $\Delta_{h_1 h_2 \dots h_m}(f)$ is the mixed finite difference of order m of the function f at the point $a = (a_1, \dots, a_m)$ determined by the increments of the variables $h_1, \dots, h_m$. A function $f \in C(\Pi(a, b))$ is called monotone if

$$L(f, \Pi(\alpha, \beta)) \geq 0 \quad \text{for all } \Pi(\alpha, \beta) \subset \Pi(a, b). \tag{3.6}$$

We denote the class of such functions by $M(\Pi(a, b))$.

An extension of the class of monotone functions. Now we introduce a weaker monotonicity condition extending the class of monotone functions. Introduce the notation

$$\begin{gathered} \overline{m} = 1, \dots, m, \qquad I_p = \{i_1, \dots, i_p\}, i_1 < \dots < i_p, \\ J_q = \{j_1, \dots, j_q\}, j_1 < \dots < j_q, \qquad I_p, J_q \subset \overline{m}, \qquad \underline{pq} = \overline{m} \backslash (I_p \cup J_q). \end{gathered} \tag{3.7}$$

Therefore I_p and J_q are ordered subsets in $\overline{m}$ containing p and q numbers, respectively. Set

$$\begin{aligned} \sum_1^x(f) &= \sum_{k=0}^{[m/2]} \sum_{I_{2k} \subset \overline{m}} L\left(f, \begin{bmatrix} b_1, \dots, x_{i_1}, \dots, x_{i_{2k}}, \dots, b_m \\ x_1, \dots, a_{i_1}, \dots, a_{i_{2k}}, \dots, x_m \end{bmatrix}\right); \\ \sum_2^x(f) &= \sum_{k=1}^{[m+1/2]} \sum_{I_{2k-1} \subset \overline{m}} L\left(f, \begin{bmatrix} b_1, \dots, x_{i_1}, \dots, x_{i_{2k-1}}, \dots, b_m \\ x_1, \dots, a_{i_1}, \dots, a_{i_{2k-1}}, \dots, x_m \end{bmatrix}\right). \end{aligned} \tag{3.8}$$

where $x = (x_1, \dots, x_m) \in \Pi(a, b)$. Let us comment on these formulas. Each one of the parallelepipeds used in (3.8) has a ("running") fixed point x as one of its vertices, while all other vertices are located on the faces of the main parallelepiped $\Pi(a, b)$ and one of them coincide with a vertex of $\Pi(a, b)$. For $k = 0$ the corresponding term in (3.8) has the form

$$L\left(f, \begin{bmatrix} b_1, \dots, b_m \\ x_1, \dots, x_m \end{bmatrix}\right).$$

We consider a class W of functions in $C(\Pi(a, b))$ for which

$$f \in W \Leftrightarrow \sum_1^x(f) \geq 0, \qquad \sum_2^x(f) \geq 0 \quad \text{for all } x \in \Pi(a, b). \tag{3.9}$$

Clearly, $M \subseteq W$.

The subspace D. The subspace D of approximants consists of sums of functions, each depending on $m - 1$ variables

$$D = \{\varphi_1(x_2, \dots, x_m) + \varphi_2(x_1, x_3, \dots, x_m) + \dots + \varphi_m(x_1, \dots, x_{m-1})\}. \tag{3.10}$$

LEMMA 3.2. *The following conditions are equivalent*:

$$(3.11)\qquad \begin{aligned} &(1) && F \in D. \\ &(2) && L(F, \Pi(\alpha, \beta)) = 0, \quad \Pi(\alpha, \beta) \subset \Pi(a, b). \\ &(3) && \sum_1^x (f) \equiv 0. \\ &(4) && \sum_2^x (f) \equiv 0. \end{aligned}$$

PROOF. The implication $1 \Rightarrow 2$ follows at once from our previous results concerning the structure of measures $\mu \in D^\perp$ (cf. Lemma 4.1 in Chapter 2, and Lemmas 1.2–1.4). Clearly, $2 \Rightarrow 3$ and $2 \Rightarrow 4$. We show that $3 \Rightarrow 1$. In parallelepipeds participating in the formula (3.8) for $\sum_1^x (F)$ the vertex $x = (x_1, \dots, x_m)$ appears $\sum_{k=0}^{[m/2]} \binom{m}{2k}$ times, and in any such parallelepiped $\delta(x) = m - 2k$, $(-1)^{\delta(x)} = (-1)^m$. Hence, $\sum_1^x (F)$ contains the term $(-1)^m/2^m \sum_{k=0}^{[m/2]} \binom{m}{2k} F(x_1, \dots, x_m)$, while each of the remaining terms depends on fewer than m variables. Hence $\sum_1^x (F) \equiv 0$ implies $F \in D$. Similarly, one can prove that $4 \Rightarrow 1$.

The best approximation of monotone functions. Let $E(f) = \operatorname{dist}(f, D)$ be the distance from $f \in C\Pi(a, b))$ to the subspace D (3.10).

THEOREM 3.3. *If $f \in W$, then*

$$E(f) = L(f, \Pi(a, b)), \tag{3.12}$$

and the function

$$\varphi(x) = f(x) - \sum_{s=0}^{m} (-1)^{s+m} \sum_{I_s \subset \overline{m}} L\left(f, \begin{bmatrix} b_1, \dots, x_{i_1}, \dots, x_{i_s} \dots, b_m \\ x_1, \dots, a_{i_1}, \dots, a_{i_s} \dots, x_m \end{bmatrix}\right) \tag{3.13}$$

gives the best approximation to f.

This theorem is due to M.-B. A. Babaev. For the proof we need the following lemmas.

LEMMA 3.4 (Additivity of L as a function on parallelepipeds). *If a parallelepiped $\Pi(a, b)$ is represented as a union of parallelepipeds $\Pi(\alpha_i, \beta_i)$,*

$$\Pi(a, b) = \bigcup_{i=1}^{s} \Pi(\alpha_i, \beta_i),$$

with pairwise disjoint interiors, then

$$L(f, \Pi(a, b)) = \sum_{i=1}^{s} L(f, \Pi(\alpha_i, \beta_i)). \tag{3.14}$$

The proof is elementary, so we omit it.

LEMMA 3.5. *The following identity holds for all* $x \in \Pi(a,b)$:

$$L(f,\Pi(a,b)) = \sum_{s=0}^{m} \sum_{I_s \subset \overline{m}} L\left(f, \begin{bmatrix} b_1,\dots,x_{i_1},\dots,x_{i_s},\dots,b_m \\ x_1,\dots,a_{i_1},\dots,a_{i_s},\dots,x_m \end{bmatrix}\right). \tag{3.15}$$

PROOF. Fix x. Any two parallelepipeds appearing on the right-hand side of (3.15) do not have common interior points, since there is at least one coordinate for which the intervals in those two parallelepipeds in which it varies have no common interior points. Moreover, the union of all those parallelepipeds is $\Pi(a,b)$, and it remains to refer to Lemma 3.4.

PROOF OF THEOREM 3.4. Let $f \in W$. Consider the function

$$F = \sum_{s=0}^{m} (-1)^{s+m} \sum_{I_s \subset \overline{m}} L\left(f \begin{bmatrix} b_1,\dots,x_{i_1},\dots,x_{i_s},\dots,b_m \\ x_1,\dots,a_{i_1},\dots,a_{i_s},\dots,x_m \end{bmatrix}\right). \tag{3.16}$$

From the proof of Lemma 3.2 ($3 \Rightarrow 1$) it follows that

$$F(x) = f(x) - \varphi(x), \qquad \varphi(x) \in D. \tag{3.17}$$

Indeed, the number of parallelepipeds in (3.16) equals $\sum_{s=0}^{m} \binom{m}{s} = 2^m$, and each of them contributes a term $2^{-m} f(x)$ to (3.17) while the remaining terms belong to D. The minus sign in front of $\varphi(x)$ is of no importance. Let us show that

$$\|F\| = L(f,\Pi(a,b)) = L(F,\Pi(a,b)). \tag{3.18}$$

The latter equality in (3.18) follows because

$$L(F,\Pi(a,b)) = L(f,\Pi(a,b)) - L(\varphi,\Pi(a,b)) = L(f,\Pi(a,b)) \tag{3.19}$$

since $\varphi \in D$ and $L \in D^{\perp}$ according to Lemma 3.2. For convenience, denote $L(f,\Pi(a,b)) = E$. In view of (3.15) we have

$$E = \sum_{1}^{x}(f) + \sum_{2}^{x}(f) \geq 0, \qquad x \in \Pi(a,b), \tag{3.20}$$

whereas from (3.16) it follows that

$$F(x) = (-1)^m \left(\sum_{1}^{x}(f) - \sum_{2}^{x}(f) \right). \tag{3.21}$$

Using (3.20) in (3.21), we obtain

$$F(x) = (-1)^m \left(E - 2\sum_{2}^{x}(f) \right), \qquad F(x) = (-1)^m \left(2\sum_{1}^{x}(f) - E \right). \tag{3.22}$$

Let m be even. Using the definition of the class W, we conclude from (3.22) that

$$-F(x) + E = 2\sum_{2}^{x}(f) \geq 0, \qquad F(x) + E = 2\sum_{1}^{x}(f) \geq 0,$$

whence $|F(x)| \leq E$. If m is odd, then (3.22) gives

$$F(x) + E = 2\sum_{2}^{x}(f) \geq 0, \qquad F(x) - E = 2\sum_{1}^{x}(f) \geq 0,$$

and again $|F(x)| \leq E$. So,

$$\|F\| \leq E = L(f, \Pi(a,b)) = L(F, \Pi(a,b)) \leq \|L\|\|F\| = \|F\|. \tag{3.23}$$

From (3.23) we obtain that

$$\|F\| = \|f - \varphi\| = L(f, \Pi(a,b)) \tag{3.24}$$

and, according to the general characterization of the best approximation, $\varphi \in D$ is the best approximation to f in D. Thus, we have proved (3.12) and (3.13) (which follows from (3.16) and (3.17)), and Theorem 3.3 follows.

Note that (3.24) implies that $\|f - \varphi\|$ is attained at all vertices of $\Pi(a,b)$ with alternation of indices in accordance with factors $(-1)^{\delta(P_i)}$ in (3.4).

An alternative expression for the best approximation. Let I_p and J_q be two intersecting subsets of $\overline{m}$. By

$$\left[(a)_{I_p}, (b)_{J_q}, x_{\underline{pq}}\right]$$

we denote a point whose coordinates with indices in I_p coincide with those of $a = (a_1, \dots, a_m)$, whose coordinates with indices in J_q coincide with those of $b = (b_1, \dots, b_m)$, and the rest of whose coordinates coincide with the coordinates of the running point $x = (x_1, \dots, x_m)$ with indices in $\overline{m} \backslash (I_p \cup J_q)$. The expression for the best approximation (3.13) now becomes

$$\varphi(x) = \sum_{p+q=1}^{m} (-1)^{p+q+1} 2^{-(p+q)} \sum_{I_p J_q \subset \overline{m}} f\left[(a)_{I_p}, b_{J_q}, x_{\underline{pq}}\right]. \tag{3.25}$$

Corollaries.

COROLLARY 3.6 (A characteristic property of the class $M(\Pi(a,b))$). *In order that $f \in M(\Pi(a,b))$, it is necessary and sufficient that for any $\Pi(\alpha,\beta) \subset \Pi(a,b)$ we have*

$$E(f)_{\Pi(\alpha,\beta)} = L(f, \Pi(\alpha,\beta)). \tag{3.26}$$

Here, $E(f)_{\Pi(\alpha,\beta)}$ is the best approximation of f by the subspace D (3.11) in the $C(\Pi(\alpha,\beta))$-metric.

PROOF. If $f \in M(\Pi(a,b))$, then, of course, $f \in M(\Pi(\alpha,\beta))$ and, in view of Theorem 3.3, (3.26) holds. If (3.26) holds, then $L(f, \Pi(\alpha,\beta)) \geq 0$ and $f \in M(\Pi(a,b))$.

THEOREM 3.7 (Flatto [48]). *Let a function f defined on $\Pi(a,b)$ have a continuous mixed partial derivative*

$$\frac{\partial^m f}{\partial x_1 \dots \partial x_m} \geq 0. \tag{3.27}$$

Then the statements of Theorem 3.3 hold.

Indeed,

$$L(f,\Pi(\alpha,\beta)) = 2^{-m} \int_{\Pi(\alpha,\beta)} \frac{\partial^m f}{\partial x_1 \dots \partial x_m} dx_1 \dots dx_m,$$

so $f \in M(\Pi(a,b))$. Theorem 3.7 generalizes Corollary 6.12 in Chapter 2.

Note in conclusion that for x_i, $i = 1, \dots, m$, in Theorem 3.7 we can take not merely coordinates but disjoint groups of coordinates. By means of the function L in [21] a two-sided estimate of $E(f)$ was also obtained for all $f \in C[a,b]$.

Approximation of products. Let D have the structure (1.11):

$$D = \sum_{j=1}^{m} H^j(\widetilde{Y}_j) \otimes C(Y_j), \tag{3.28}$$

where the sets $\widetilde{Y}_j$ are pairwise disjoint (the H^j are still finite dimensional subspaces in $C(\widetilde{Y}_j)$).

THEOREM 3.8. *Let*

$$f = \prod_1^m f_j\left(\tilde{y}_j\right), \qquad f_j\left(\tilde{y}_j\right) \in C(\widetilde{Y}_j), \quad j = 1, \dots, m, \tag{3.29}$$

and let the $\varphi_j\left(\tilde{y}_j\right) \in H^j$ be the best approximations to f_j in $C(\widetilde{Y}_j)$ by subspaces H_j, whereas $E^j = \|f_j - \varphi_j\|$ is the magnitude of that approximation for $j = 1, \dots, m$. Let $E(f)$ be the magnitude of the best approximation of the function f by the subspace D in $C(Y)$. Then

$$E(f) = \prod_{j=1}^m E_j, \tag{3.30}$$

and

$$\varphi = f - \prod_1^m (f_j - \varphi_j) \in D \tag{3.31}$$

is the best approximation to f.

PROOF. First of all, it is clear that φ indeed belongs to D. Furthermore, according to the general criterion for the best approximation (Theorem 6.2 in Chapter 2) there exist linear functionals $\lambda_j \in H^{j\perp} \subset C\left(Y_j\right)$ with the following properties:

$$\|\lambda_j\| = 1, \quad \lambda_j\left(f_j\right) = \|f_j - \varphi_j\|_{C(\widetilde{Y}_j)} = E^j, \qquad j = 1, \dots, m.$$

The functional $\lambda = \lambda_1 \otimes \cdots \otimes \lambda_m$ belongs to $C(Y)^\star$, $\|\lambda\| = 1$, and by Lemma 1.5, $\lambda \in D^\perp$. We have

$$\begin{aligned} 0 =&\lambda(\varphi) = \lambda[f] - \prod_1^m \lambda_j\,[f_j - \varphi_j] \\ =&\lambda[f] - \prod_1^m \|f_j - \varphi_j\|_{C(\tilde{Y}_j)} = \lambda(f) - \|f - \varphi\|_{C(Y)}, \end{aligned}$$

i.e.,

$$\lambda[f] = \|f - \varphi\|_{C(Y)}. \tag{3.32}$$

From (3.32) all statements of the theorem follow.

Theorem 3.8 is due to Shapiro [**118a**]. A more abstract version can be found in [**24**].

A problem in L^2. Digressing from our earlier agreement not to consider metrics other than C and ℓ^∞, we now consider a problem of approximation by sums in the L^2-metric. The reason is that a beautiful result obtained in that problem stems from the Boolean properties of projections closely related to the approach developed in §1 of this chapter. Only this time we are speaking about *orthogonal* projections—a natural object in the Hilbert space theory (see, e.g., [**118**], [**40**]). The above-mentioned result also deserves our attention here because it was probably the first for which an explicit formula for the solution was obtain in a rather general problem of approximation by sums of functions of a smaller number of variables.

Let $X_1, \dots, X_N$ be sets, and on each X_i, $i = 1, \dots, N$, let there be a σ-algebra of subsets $\mathcal{M}_i$ on which there is defined a probability measure μ_i:

$$\mu_i \geq 0, \qquad \int_{X_i} d\mu_i = 1. \tag{3.33}$$

So, $(X_i, \mathcal{M}_i, \mu_i)$ are measure spaces with probability measures. If $Y = X_1 \times \cdots \times X_N$, then we define a measure on Y by setting

$$\mu = \mu_1 \times \cdots \times \mu_N. \tag{3.34}$$

If $(j) = (i_1, \dots, i_N)$ is a subset of indices in $\overline{N} = \{1, \dots, N\}$, $Y_{(j)} = X_{i_1} \times \cdots \times X_{i_j}$, then there is an induced measure on $Y_{(j)}$

$$\mu_{(j)} = \mu_{i_1} \times \cdots \times \mu_{i_j}. \tag{3.35}$$

(When considering different products we keep the same conventions about the order as in §1.) As in §1, $(\tilde{j})$ denotes the set of complementary indices to (j). The spaces $L^2(Y_{(j)}, \mu_{(j)})$ can be naturally embedded as subspaces into $L^2(Y, \mu)$. For a function $f \in L^2(y, \mu)$, $f(y_{(j)})$ denotes its mean over $\widetilde{Y}_{(j)}$:

$$f\left(y_{(j)}\right) = \int_{Y_{(\tilde{j})}} f d\mu_{(\tilde{j})}. \tag{3.36}$$

A direct calculation (here and below everything is based on the Fubini theorem) yields:

LEMMA 3.9. *The formula* (3.36) *defines an orthogonal projection* $P_j : L^2(Y,\mu) \to L^2(Y_{(j)}, \mu_{(j)})$. *The complementary orthogonal projection* R_j *is defined by the formula* $f - f\left(y_{(j)}\right)$.

LEMMA 3.10. *Let* E *be a Hilbert space,* $E_1, \dots, E_m \subset E$ *closed subspaces,* $P_k : E \to E_k$ *orthogonal projections, and* R_k *orthogonal projections complementary to* P_k, $k = 1, \dots, m$. *If the projections* P_k, $k = 1, \dots, m$, *commute pairwise, then* (1.43) *defines an orthogonal projection* P *of* E *onto* $E_1 + \cdots + E_m$, *while* (1.44) *defines the complementary projection.*

The proof is obtained by induction, making use of elementary properties of orthogonal projections.

LEMMA 3.11. *Let* (j) *and* (k) *be two sets of indices in* $\overline{N}$ *and, moreover,* $(j) \not\subset (k)$, $(k) \not\subset (j)$. *Then, the projections* P_j *and* P_k *defined by the formula* (4.46) *commute, and for* $f \in L^2(Y,\mu)$ *we have*

$$P_k P_j(f) = P_j P_k(f) = f\left(y_{(j)\cap(k)}\right) = \int_{Y(\tilde{j})\cup(\tilde{k})} f d\mu_{(\tilde{j})\cup(k)}. \tag{3.37}$$

In particular, if $(j) \cap (k) = \emptyset$, *then* $f\left(y_{(j)\cap(k)}\right) = \text{const} = \int_Y f d\mu.$

Let $(j_1), \dots, (j_m)$ be subsets in $\tilde{N}$ such that $(j_k) \not\subset (j_\ell)$ when $k \neq \ell$, $k, \ell = 1, \dots, m$. Consider in $L^2(Y,\mu)$ the problem of best approximation to a function $f(y) \in L^2(Y,\mu)$ by functions from the subspace

$$L^2 D = \left\{\varphi_1\left(y_{(j_1)}\right) + \cdots + \varphi_m\left(y_{(j_m)}\right)\right\}. \tag{3.38}$$

THEOREM 3.12 (Mordashev [**106–107**]). *The unique function in the subspace* $L^2 D$ *with the least deviation from a function* $f \in L^2(Y,\mu)$ *is given by*

$$\begin{aligned} F(y) = {} & \sum_{1\le k\le m} f\left(y_{(j_k)}\right) - \sum_{1\le k\le \ell\le m} f\left(y_{(j_k)\cap(j_\ell)}\right) \\ & + \sum_{1\le k\le \ell\le r\le m} f\left(y_{(j_k)\cap(j_\ell)\cap(j_r)}\right) + \cdots + (-1)^{m-1} f\left(y_{(j_1)\cap(j_2)\cap\cdots\cap(j_m)}\right). \end{aligned} \tag{3.39}$$

A proof follows at once from the fact that the function least deviating from f must be an orthogonal projection of f on the approximating subspace (this follows from well-known properties of Hilbert spaces) and, according to preceding lemmas, must be of the form (3.39). Uniqueness of the best approximation in a Hilbert space is also well known.

Let $d\mu_1 = d\mu_2 = dx = dy$ be the ordinary Lebesgue measure on $I = [0,1]$. For $L^2\left(I^2, dx \otimes dy\right)$ consider the problem of best approximation by functions $\varphi(x) + \psi(y)$.

COROLLARY 3.13 (Denisyuk [**37**]). *The function least deviating from a given function* $f(x,y)$ *is unique and is given by the formula*

$$F = \int_I f(x,y)dy + \int_I f(x,y)dx - \int_{I^2} f dx dy.$$

In [**69**], [**70**], certain problems of the best approximation in the L^1-metric were more effectively solved by using ideas similar to those presented above.

On the Diliberto–Straus algorithm when the number of variables is greater than 2. The levelling algorithm of Diliberto and Straus studied in §7 of Chapter 2 gives an effective method of constructing the best approximation of a function of two variables by sums $\varphi(x)+\psi(y)$. In the case of three or more variables it may not succeed. This was first discovered by Aumann [**13**], whose paper does not appear to be widely known, and was repeated in [**104**].

A counterexample to the algorithm in $\mathbb{R}^3$ ([104]). For the sake of clarity, first consider the problem of approximating a continuous function $f(x_1, x_2, x_3)$ on the unit cube I^3, $I = [0,1]$, by sums $\varphi(x_1)+\psi(x_2)+\chi(x_3)$. Our goal is to construct a function $f(x_1, x_2, x_3)$ for which a natural extension of the levelling algorithm does not lead to the construction of a best approximation. Define a continuous function on $[0,3]$ by setting

$$F(t) = \begin{cases} -1, & 0 \le t \le 1; \\ \text{linear}, & 1 \le t \le 2; \\ 1, & 2 \le t \le 3 \end{cases}$$

and let $f(x_1, x_2, x_3) = F(x_1 + x_2 + x_3)$ on I^3 ($\|f\| = 1$). Also, define a function $\hat{F}(t) = F(t) - (t - 3/2)$. Then

$$\hat{f}(x_1, x_2, x_3) = \hat{F}(x_1 + x_2 + x_3) = f(x_1, x_2, x_3) - [(x_1 - 3/2) + (x_2 + x_3)].$$

Let us show that the sum

$$\varphi(x_1) + \psi(x_2) + \chi(x_3) = (x_1 - 3/2) + x_2 + x_3$$

is the best approximation to $f(x_1, x_2, x_3)$ among all sums $\varphi(x_1) + \psi(x_2) + \chi(x_3)$. Note that $\|\hat{F}\| = 1$, and if we construct the functional $L(f, I^3)$ by (3.4), then

$$L(\hat{f}, I^3) = \frac{1}{2} = \|\hat{f}\|$$

which proves the above assertion, since $L \in \{\varphi + \psi + \chi\}^{\perp}$. (The values of f at the vertices of the cube equal $\pm\frac{1}{2}$, and the signs coincide with those of the charges at these vertices.) The function f takes the value -1 on all the edges emanating from the vertex $(0,0,0)$ and the value 1 on the edges emanating from the vertex $(1,1,1)$. Therefore, on any cross-section of I^3 by a plane $x_1 = x_1^0$ its maximum equals 1, while the minimum equals -1. Therefore, it is levelled with respect to the variables (x_2, x_3). (Similarly, it is levelled with respect to (x_1, x_2) and (x_2, x_3).) Starting the levelling process, we have to take

$$\varphi(x_1) - \frac{1}{2}\left[\max_{(x_2,x_3)} f(x_1, x_2, x_3) + \min_{(x_2,x_3)} f(x_1, x_2, x_3)\right] = 0, \qquad f - \varphi = f,$$

and, similarly, $\psi(x_2) = 0$ and $\chi(x_3) = 0$. So, the levelling process will not move away from the already-levelled function f and we will not approach the magnitude of the best approximation, equal to $\frac{1}{2}$.

A counterexample to the algorithm in $\mathbb{R}^N$ [104]. Consider an extension of the above example to I^N. Consider on $[0, N]$ a continuous function

$$F(t) = \begin{cases} -1, & 0 \le t \le 1, \\ \text{a linear function } \left(= \frac{2t-N}{N-2}\right), & 1 \le t \le N-1, \\ 1, & N-1 \le t \le N \end{cases}$$

and

$$\hat{F}(t) = F(t) - \frac{2t - N}{N-1}, \qquad \|\hat{F}\| = \frac{1}{N-1}.$$

Set

$$f(x_1, \dots, x_N) = F(x_1 + \dots + x_N);$$
$$\hat{f}(x_1, \dots, x_N) = \hat{F}(x_1 + \dots + x_N).$$

Clearly,

$$\hat{f} = f - \sum_1^N g_i(x_i)$$

and we show that $\hat{f}$ has the least norm among all functions that have such a form ($\|\hat{f}\| = \frac{1}{N-1}$). To do that, construct a functional L with the following properties:

$$\|L\| = 1, \quad L \in \{\varphi_1(x_1) + \dots + \varphi_N(x_N)\}^\perp, \qquad L(\hat{f}) = \|\hat{f}\|. \tag{3.40}$$

To construct L, place nonnegative charges $a, a_1, \dots, a_N$ at the vertices

$$(0, 0, \dots, 0), (0, 1, \dots, 1), (1, 0, 1, \dots, 1), \dots, (1, 1, \dots, 1, 0)$$

(at these vertices $\hat{f}$ assumes its maximum $\frac{1}{N-1}$) and nonpositive charges $-b, -b_1, \dots, -b_N$ at the vertices

$$(1, \dots, 1), (1, 0, \dots, 0), (0, 1, 0, \dots, 0), \dots, (0, 0, \dots, 1),$$

where $\hat{f}$ assumes its minimum (equal to $-\frac{1}{N-1}$). To satisfy the requirements in (3.40) we set

$$a = b = \frac{N-2}{4(N-1)}, \qquad a_i = b_i = \frac{1}{4(N-1)}. \tag{3.41}$$

For the functional L corresponding to these charges we obtain

$$\|L\| = a + b + \sum_1^N (a_i + b_i) = \frac{N-2}{2(N-1)} + \frac{N}{2(N-1)} = 1.$$

The second requirement in (3.40) holds, provided that in every hyperplane $x_i =$ const, $i = 1, \dots, N$, the total charge equals zero (cf. Lemma 4.1 in Chapter 2 and Lemmas 1.2–1.4 in this chapter). There are no charges in a hyperplane $x_i = c$, $0 < c < 1$. Calculate the charge in the hyperplane $x_i = 0$. In view of (3.41) it equals

$$a + a_i - \sum_{j \neq i, j=1}^N b_j = \frac{N-2}{4(N-1)} + \frac{1}{4(N-1)} - \frac{N-1}{4(N-1)} = 0.$$

Similarly, in the hyperplane $x_i = 1$ we have

$$-b - b_i + \sum_{j \neq i, j=1}^N a_j = 0.$$

So, $L \in \{\varphi_1(x_1) + \dots + \varphi_N(x_N)\}^\perp$. Finally, in view of the choice of signs of charges in L we obtain

$$L(\hat{f}) = \frac{1}{N-1} = \|\hat{f}\|.$$

However, for the same reasons as in the case $N = 3$ the levelling algorithm for the function f cannot change that function. Hence, it does not lead to finding the magnitude of the best approximation.

Discussion. From the above example it follows that for a function f to be levelled with respect to all groups of $(N-1)$ variables, it is not sufficient that f have the minimal norm among among all functions $f - (g_1(x_1) + \cdots + g_N(x_N))$ as was the case for $N = 2$ (cf. §6 in Chapter 2). Nevertheless, it would be interesting to know whether every function can be levelled. In other words, we ask if an arbitrary function f can be represented in the form

$$f(x_1, \dots, x_N) = F(x_1, \dots, x_N) + \sum_1^N g_i(x_i),$$

where F is a corresponding levelled function. In Aumann's paper [**13**] it is also shown that even for approximation in $C(I^3)$ by functions

$$\varphi(x_1, x_2) + \psi(x_1, x_3) + \chi(x_2, x_3)$$

the levelling algorithm need not be successful.

References

1. R. A. Aleksandryan and E. A. Mirzahanyan, *General topology*, "Vysshaya Shkola", Moscow, 1979. (Russian)
2. N. I. Akhiezer, *Lectures on approximation theory*, 2nd rev. ed., "Nauka", Moscow, 1965; English transl. of 1st ed., Ungar, New York, 1956; German transl. of 2nd ed., Akademie-Verlag, Berlin, 1967.
3. A. A. Anikevitch and A. B. Gribov, *The approximation of elements of a matrix by the sum of corresponding components of two vectors*, Operations Research and Statistical Simulation, Vyp. 1, Izdat. Leningrad. Univ., Leningrad, 1972, pp. 3–9. (Russian)
4. V. I. Arnold, *On functions of three variables*, Dokl. Akad. Nauk SSSR **114** (1957), 679–681; English transl, Amer. Math. Soc. Transl. (2) **38** (1963), 51–54.
5. ———, *On the representation of continuous functions of three variables by superpositions of continuous functions of two variables*, Mat. Sb. **48** (1959), 3–74; English transl., Amer. Math. Soc. Transl. (2) **28** (1963), 61–147.
6. ———, *Some questions of approximation and representation of functions*, Proc. Internat. Congr. Math. (Edinburgh, 1958), Cambridge Univ. Press, Cambridge, 1960, pp. 339–348; English transl., Amer. Math. Soc. Transl. (2) **53** (1966), 192–201.
7. ———, *On the representability of functions of two variables in the form* $\chi(\varphi(x) + \psi(y))$, Uspekhi Mat. Nauk **12** (1957), no. 2, 119–121. (Russian)
8. ———, *On representability of functions of several variables in the form of superpositions of functions of fewer variables*, Mat. Prosveshchenie **1958**, no. 3, 41–61. (Russian)
9. G. Aumann, *Über approximative Nomographie.* I, Bayer. Akad. Wiss. Math.-Nat. Kl. S.-B. **1958**, 137–155.
10. ———, *Über approximative Nomographie.* II, Bayer. Akad. Wiss. Math.-Nat. Kl. S.-B. **1959**, 103–109.
11. ———, *Über approximative Nomographie.* III, Bayer. Akad. Wiss. Math.-Nat. Kl. S.-B. **1960**, 27–34.
12. ———, *Linear Approximationen auf einem Gefled*, Arch. Math. (Basel) **10** (1959), 267–272.
13. ———, *Approximation by step functions*, Proc. Amer. Math. Soc. **14** (1963), 477–482.
14. ———, *Approximation von Funktionen.* I: *Theoretische Grundlagen*, Mathematische Hilfsmittel des Ingenieurs. Vol. III (R. Sauer and I. Szabo, eds.), Springer-Verlag, Berlin, 1968, pp. 320–446.
15. M.-B. A. Babaev, *Approximation to functions of several variables by functions of a smaller number of variables*, Approximation and Function Spaces (Z. Ciesielski, ed.), North Holland, Amsterdam, 1981, pp. 44–50.
16. ———, *Approximation of polynomials of two variables by sums of functions of one variable*, Izv. Akad. Nauk Azerbaidzhan. SSR Ser. Fiz-Tekhn. Mat. Nauk **1971**, no. 2, 23–29. (Russian)
17. ———, *Approximation of functions of several variables by sums of functions of fewer variables in the complex domain*, Special Questions on Differential Equations and Function Theory, "Èlm", Baku, 1970, pp. 3–44. (Russian)
18. ———, *Approximation of polynomials of two variables by functions of the form* $\varphi(x)+\psi(y)$, Dokl. Akad. Nauk SSSR **193** (1970), 967–969; English transl., Soviet Math. Dokl. **11** (1970), 1034–1036.

19. ______, *Extremal elements and the value of the best approximation of a monotone function in $\mathbb{R}^n$ by sums of functions of fewer variables*, Dokl. Akad. Nauk SSSR **265** (1982), 11–13; English transl., Soviet Math. Dokl. **26** (1982), 1–4.
20. ______, *Best approximation by functions of fewer variables*, Dokl. Akad. Nauk SSSR **279** (1984), 273–277; English transl., Soviet Math. Dokl. **30** (1984), 629–632.
21. ______, *Extremal properties and two-sided estimates for approximation by sums of functions of fewer variables*, Mat. Zametki **36** (1984), 647–659; English transl., Math. Notes **36** (1984), 821–828.
22. L. A. Bassalygo, *On the representations of continuous functions of two variables by means of continuous functions of one variable*, Vestnik Moskov. Univ. Ser. I Mat. Mekh. **1966**, no. 1, 58–63. (Russian)
23. T. E. Bogatkina and V. A. Dozorov, *Approximation of a function of two variables by a sum of functions of linear combination of arguments*, Avtomat. i Telemekh. **1968**, no. 11, 156–159; English transl., Automat. Remote Control **29** (1968), 1861–1863.
24. Yu. A. Brudnyi, *Approximation of functions of n variables by quasi-polynomials*, Izv. Akad. Nauk SSSR Ser. Mat. **34** (1970), 564–583; English transl., Math. USSR Izv. **4** (1970), 568–586.
25. R. C. Buck, *Approximate complexity and functional representation*, J. Math Anal. Appl. **70** (1979), 280–298.
26. ______, *Approximation theory and functional equations*, J. Approx. Theory **5** (1972), 228–237.
27. ______, *Approximation theory and functional equations.* II, J. Approx. Theory **9** (1973), 121–125.
28. ______, *On the functional equation* $\varphi(x) = g(x)\Phi(B(x)) + h$, Proc. Amer. Math. Soc. **31** (1972), 159–161.
29. A. S. Cereteli, *Approximation of functions of several variables by functions of the form* $\Phi_1(x_1) + \cdots + \Phi_n(x_n)$, Tbilisi Sahema Univ. Srom. Mekh.–Mat. Ser. **129** (1968), 397–409. (Russian)
30. E. W. Cheney, *The best approximation of multivariate functions by combinations of univariate ones*, Approximation Theory. IV (C. Chui, ed.), Academic Press, New York, 1983, pp. 1–26.
31. ______, *Recent progress in multivariate approximation*, Constructive Theory of Functions (Proc. Internat. Conf., Varna, 1984; Bl. Sendov et al., eds.), Publ. House Bulgar. Acad. Sci., Sofia, 1984, pp. 208–213.
32. ______, *Algorithms for approximation*, Approximation Theory, Proc. Sympos. Appl. Math., vol. 36, Amer. Math. Soc., Providence, RI, 1986, pp. 67–80.
33. L. Ciobanu, *Approximation of continuous functions of two variables* $f(x, y)$ *by polynomials* $P(z)$, *where* $z = xy$, Bull. Inst. Polytech. Iaşi (N. S.) **13(17)** (1967), 135–138. (Russian)
34. L. Collatz, *Approximation by functions of fewer variables*, Conf. Theory of Ordinary and Partial Differential Equations (Dundee, 1972), Lecture Notes in Math., vol. 280, Springer-Verlag, Berlin, 1972, pp. 16–21.
35. L. Collatz and W. Krabs, *Approximationstheorie. Tschebyscheffsche Approximationen mit Anwendungen*, Teubrer, Stuttgart, 1973.
36. M. M. Day, *Normed linear spaces*, 2nd ed., Springer-Verlag, Berlin, 1973.
37. I. N. Denisyuk, *Analytic methods of approximate correlation and corresponding functional problems*, Moskov. Gos. Univ. Uchen. Zap. **28** (1939), 27–42. (Russian)
38. S. P. Diliberto and E. G. Straus, *On the approximation of a function of several variables by the sum of functions of fewer variables*, Pacific J. Math. **1** (1951), 195–210.
39. R. Doss, *On the representation of continuous functions of two variables by means of addition and continuous functions of one variable*, Colloq. Math. **10** (1963), 249–259.
40. N. Dunford and J. T. Schwartz, *Linear operators.* Part I, Interscience, New York, 1959.
41. N. Dyn, *A straightforward generalization of Diliberto and Straus' algorithm does not work*, J. Approx. Theory **30** (1980), 247–250.
42. R. E. Edwards, *Functional analysis: theory and applications*, Holt, Rinehart and Winston, New York, 1965.
43. V. Ya. Èĭderman, *On the levelling algorithm for approximation of a bivariate function by sums of univariate functions*, Voprosy Mat. i Mekh. Sploshn. Sred, Moscow Inst. Civil Engrg., Moscow, 1984, pp. 67–84. (Russian)

44. ———, *On the algorithm of Diliberto and Straus for approximation of bivariate functions by sums* $g(x) + h(y)$, Sibirsk Mat. Zh. **28** (1987), no. 5, 223–224. (Russian)
45. ———, *On the algorithm of Diliberto and Straus for approximation of bivariate functions by sums of univariate functions*, Voporsy Mat. i Mekh. Sploshn. Sred, Moscow Inst. Civil Engrg., Moscow, 1987, pp. 92–96. (Russian)
45a. L. I. Epstein, *Nomography*, Interscience, New York, 1958.
46. Guohui Feng and M. von Golitschek, *On the failure of proximality of tensor-product subspaces*, J. Approx. Theory **62** (1990), 340–349.
47. S. D. Fisher, *The decomposition of* $C_r(K)$ *into the direct sum of subalgebras*, J. Funct. Anal. **31** (1979), 218–273.
48. L. Flatto, *The approximation of certain functions of several variables by sums of functions of fewer variables*, Amer. Math. Monthly **74** (1966), 131–132.
49. C. Franchetti and E. W. Cheney, *Minimal projections in tensor-product spaces*, J. Approx. Theory **41** (1984), 367–381.
50. ———, *Simultaneous approximation and restricted Chebyshev centers in function spaces*, Approximation Theory and Applications (Z. Ziegler, ed.), Academic Press, New York, 1981, pp. 65–88.
51. ———, *The embedding of proximinal sets*, J. Approx. Theory **40** (1986), 213–225.
52. B. L. Fridman, *An improvement in the smoothness of function in the Kolmogorov superposition theory*, Dokl. Akad. Nauk SSSR **177** (1967), 1019–1022; English transl., Soviet Math. Dokl. **8** (1967), 1550–1553.
53. D. R. Fulkerson and P. Wolfe, *An algorithm for scaling matrices*, SIAM Rev. **4** (1962), 142–146.
54. A. L. Garkavi, S. Ya. Khavinson, and V. A. Medvedev, *On the existence of the best uniform approximation of a function of two variables by the sums* $\varphi(x) + \psi(y)$, Sibirsk. Mat. Zh. **36** (1995), 819–827; English transl., Siberian Math. J. **36** (1995), 707–713.
54a. ———, *On the existence of the best uniform approximation of a function of several variables by sums of functions of fewer variables* (to appear).
55. J. B. Garnett, *Bounded analytic functions*, Academic Press, New York, 1981.
56. M. von Golitschek, *An algorithm for scaling matrices and computing the minimum cycle mean in a digraph*, Numer. Math. **35** (1980), 45–55.
57. ———, *Approximating bivariate functions and matrices by nomographic functions*, Quantitative Approximation (R. de Vore and K. Scherer, eds.), Academic Press, New York, 1980, pp. 143–151.
58. ———, *Remarks on functional representation*, Approximation Theory. III (E. W. Cheney, ed.), Academic Press, New York, 1980, pp. 429–434.
59. ———, *Approximation of functions of two variables by the sum of two functions of one variable*, Numerical Methods of Approximation Theory (L. Collatz et al., eds.), Birkhäuser, Basel, 1980, pp. 117–124.
60. ———, *On the existence of continuous best approximations in tensor-product subspaces of univariate functions*, Approximation Theory. IV (C. Chui, ed.), Academic Press, New York, 1983, pp. 475–481.
61. M. von Golitschek and E. W. Cheney, *On the algorithm of Diliberto and Straus for approximating bivariate functions by univariate ones*, Numer. Funct. Anal. Optim. **1** (1979), 341–363.
62. ———, *The best approximation of bivariate functions by separable functions*, Topological Methods in Nonlinear Functional Analysis (S. P. Singh et al., eds), Contemporary Math., vol. 21, Amer. Math. Soc., Providence, RI, 1983, pp. 125–135.
63. ———, *Failure of the alternating algorithm for best approximation of multivariate functions*, J. Approx. Theory **38** (1983), 139–143.
64. M. Golomb, *Approximation by functions of fewer variables*, On Numerical Approximation (R. Langer, ed.), University of Wisconsin Press, Madison, WI, 1959, pp. 275–327.
65. H. M. Gonska, *Degree of simultaneous approximation of bivariate functions by Gordon operators*, J. Approx. Theory **62** (1990), 171–191.
66. W. Y. Gordon, *Blending-function method of bivariate and multivariate interpolation and approximation*, SIAM J. Numer. Anal. **8** (1971), 158–177.

67. ———, *Distributive lattices and the approximation of multivariate functions*, Approximation with Special Emphasis on Spline Functions (J. Schoenber, ed.), Academic Press, New York, 1969, pp. 223–277.
68. W. Y. Gordon and E.W. Cheney, *Bivariate and multivariate interpolation with noncommutative projectors*, Linear Spaces and Approximation (P. L. Butzer and B. Sz.-Nagy., eds.), Birkhäuser, Basel, 1978, pp. 381–387.
69. W. Haussman and K. Zeller, *Uniqueness and non-uniqueness in bivariate L^1-approximation*, Approximation Theory. IV (C. Chui, ed.), Academic Press, New York, 1983, pp. 509–514.
70. ———, *Blending interpolation and best L^1-approximation*, Arch. Math. (Basel) **40** (1983), 545–552.
71. T. Hedberg, *The Kolmogorov superposition theorem*, Appendix II to H. S. Shapiro, *Topics in Approximation Theory*, Lecture Notes in Math., vol. **187**, Springer-Verlag, Berlin, 1971, pp. 267–275.
72. D. Hilbert, *Mathematische Probleme*, Nachr. Akad. Wiss. Göttingen **1900**, 253–297; reprinted in his *Gesammelte Abhandlungen*, Vol. 3, Springer-Verlag, Berlin, 1935, pp. 290–329; English transl., Bull. Amer. Math. Soc. **8** (1902), 461–462; reprinted in Mathematical Developments Arising from Hilbert Problems, Proc. Sympos. Pure Math., vol. 28, Amer. Math. Soc., Providence, RI, 1976, pp. 1–34
72a. W. Hurewicz and H. Wallman, *Dimension theory*, Princeton Univ. Press, Princeton, NJ, 1948.
73. J.-P. Kahane, *Séries de Fourier absolument convergentes*, Springer-Verlag, Berlin, 1970.
74. ———, *Sur le théorème de superposition de Kolmogorov*, J. Approx. Theory **13** (1975), 229–234.
75. V. A. Kaminskiĭ and V. I. Makarov, *On the problem of the conditional levelling on a discrete grid*, Voprosy Mat. i Mekh. Sploshn. Sred, Moscow Inst. Civil Engrg., Moscow, 1979, pp. 137–142. (Russian)
76. M. Katětov, *On real functions in topological spaces*, Fund. Math. **38** (1951), 85–91.
77. C. T. Kelley, *A note on the approximation of functions of several variables by sums of functions of one variable*, J. Approx. Theory **33** (1981), 179–189.
78. J. L. Kelley, *General topology*, Van Nostrand, New York, 1955.
79. S. Ya. Khavinson [Havinson], *A Chebyshev theorem for the approximation of a function of two variables by sums of the type $\varphi(x)+\psi(y)$*, Izv. Akad. Nauk SSSR Ser. Mat. **33** (1969), 650–666; English transl., Math. USSR Izv. **3** (1969), 617–632.
80. ———, *On the representation of functions of two variables by the sums $\varphi(x)+\psi(y)$*, Izv. Vyssh. Uchebn. Zaved. Mat. **1985**, no. 2, 66–73; English transl., Soviet Math. (Iz. VUZ) **29** (1985), no. 2, 81–90.
81. ———, *On representation and approximation of functions of two variables by linear superpositions*, Voprosy Mat. i Mekh. Sploshn. Sred, Moscow Inst. Civil Engrg., Moscow, 1987, pp. 89–92. (Russian)
82. ———, *The annihilator of linear superpositions*, Algebra i Analiz **7** (1995), no. 3, 1–42; English transl., St.-Petersburg Math. J. **7** (1996), 307–341.
83. ———, *Certain approximate properties of linear superpositions*, Izv. Vyssh. Uchebn. Zaved. Mat. **1995**, no. 8, 63–73; English transl., Russian Math. (Iz. VUZ) **39** (1995), no. 8, 60–70.
84. ———, *On interpolational properties of outer functions in Kolmogorov's superpositions*, Mat. Zametki **55** (1994), no. 4, 104–113; English transl., Math. Notes **55** (1994), 406–412.
85. A. N. Kolmogorov, *On the representation of continuous functions of several variables by superpositions of continuous functions of fewer variables*, Dokl. Akad. Nauk SSSR **108** (1956), 179–182; English transl., Amer. Math. Soc. Transl. (2) **17** (1961), 369–373.
86. ———, *On the representation of continuous functions of several variables by superpositions of continuous functions of one variable and addition*, Dokl. Akad. Nauk SSSR **114** (1957), 953–956; English transl., Amer. Math. Soc. Transl. (2) **28** (1963), 55–59.
87. A. N. Kolmogorov and V. M. Tikhomirov, *ε-entropy and ε-capacity of sets in function spaces*, Uspekhi Mat. Nauk **13** (1959), no. 2, 3–86; English transl., Amer. Math. Soc. Transl. (2) **17** (1961), 277–364.
88. S. V. Konyagin, *On proximinality of some sets in L^p, $1 \le p \le \infty$*, Abstracts Second Internat. Conf. Functional Analysis and Approximation Theory, Univ. Basilicata, Potenza, Italy, 1992.

89. M. G. Kreĭn and A. A. Nudel′man, *The Markov moment problem and extremal problems*, "Nauka", Moscow, 1973; English transl., Amer. Math. Soc., Providence, RI, 1977.

90. B. K. Kripke and K. B. Holmes, *Approximation of bounded functions by continuous functions*, Bull. Amer. Math. Soc. **71** (1965), 896–897.

90a. M. Kuczma, *Functional equations in a single variable*, PWN, Warsaw, 1968.

91. W. A. Light and E. W. Cheney, *On the approximation of a bivariate function by the sum of univariate functions*, J. Approx. Theory **29** (1980), 305–322.

92. ______, *Some best approximation theorems in tensor-product spaces*, Math. Proc. Cambridge Philos. Soc. **89** (1981), 385–390.

93. ______, *The characterization of best approximation of tensor-product spaces*, Analysis **4** (1984), 1–26.

94. ______, *Approximation theory in tensor-product spaces*, Lecture Notes in Math., vol. 1169, Springer-Verlag, Berlin, 1985.

95. G. G. Lorentz, *Metric entropy, widths and superpositions of functions*, Amer. Math. Monthly **69** (1962), 469–485.

96. ______, *Approximation of functions*, Holt, Rinehart and Winston, New York, 1966.

97. ______, *The* 13*th problem of Hilbert*, Mathematical Developments Arising from Hilbert Problems, Proc. Sympos. Pure Math., vol. 28, Amer. Math. Soc., Providence, RI, 1976, pp. 419–429.

98. ______, *Superpositions, metric entropy, complexity of functions, widths*, Bull. London Math. Soc. **22** (1990), 64–71.

99. D. E. Marshall and A. G. O'Farrell, *Uniform approximation of real functions*, Fund. Math. **104** (1979), 203–211.

100. ______, *Approximation by a sum of two algebras, the lightning bolt principle*, J. Funct. Anal. **52** (1983), 353–368.

101. V. A. Medvedev, *On optimal representation of functions by sum of two compositions*, Izv. Vyssh. Uchebn. Zaved. Mat. **1991**, no. 5, 41–49; English transl., Soviet Math. (Iz. VUZ) **35** (1991), no. 5, 42–51.

102. ______, *On the sum of two closed algebras of continuous functions on a compact*, Funktsional. Anal. i Prilozhen. **27** (1991), no. 1, 33–36; English transl., Functional Anal. Appl. **27** (1991), 28–30.

103. ______, *On representation of functions by a sum of several compositions*, Mat. Sb. **182** (1991), 1379–1392; English transl., Math. USSR Sb. **74** (1993), 119–130.

104. ______, *Refutation of a theorem of Diliberto and Straus*, Mat. Zametki **51** (1992), no. 4, 79–80; English transl., Math. Notes **51** (1992), 380–381.

105. E. Michael, *Continuous selections*, Ann. of Math. (2) **63** (1956), 361–382.

106. ______, *Selected selection theorems*, Amer. Math. Monthly **63** (1956), 233–238.

107. V. M. Mordashev, *Approximation of a functions of several variables by a sum of functions of fewer variables*, Dokl. Akad. Nauk SSSR **183** (1968), 778–779; English transl., Soviet Math. Dokl. **9** (1968), 1462–1463.

108. ______, *The best approximation of a function of several variables by a sum of functions of fewer variables*, Mat. Zametki **5** (1969), 217–226; English transl., Math. Notes **5** (1969), 132–137.

109. V. P. Motornyi, *On a problem of the best approximation of functions of two variables by functions of the form* $\varphi(x) + \psi(y)$, Izv. Akad. Nauk SSSR Ser. Mat. **27** (1963), 1211–1214. (Russian)

110. ______, *On a problem of the best approximation of functions of two variables by functions of the form* $\varphi(x) + \psi(y)$, Studies in Modern Constructive Function Theory (Proc. Second All-Union Conf., Baku, 1962), "Èlm", Baku, 1965, pp. 66–72. (Russian)

111. I. P. Natanson, *Theory of functions of a real variable*, 3rd ed., "Nauka", Moscow, 1974; English transl., Vols. I (1st ed., Chaps. I–IX), II (2nd ed., Chaps. X–XVI, XVIII), Ungar, New York, 1955, 1961.

112. A. G. O'Farrell, *Five generalizations of the Weierstrass approximation theorem*, Proc. Royal Irish Acad. **81**A (1981), 65–69.

113. Yu. P. Ofman, *On the best approximation of functions of two variables by functions of the form* $\varphi(x) + \psi(y)$, Izv. Akad. Nauk SSSR Ser. Mat. **25** (1961), 239–252; English transl., Amer. Math. Soc. Transl. (2) **44** (1965), 12–29.

114. P. A. Ostrand, *Dimension of metric spaces and Hilbert's problem* 13, Bull. Amer. Math. Soc. **71** (1965), 619–622.
115. J. R. Respess and E. W. Cheney, *Best approximation problems in tensor-product spaces*, Pacific J. Math. **102** (1982), 437–446.
116. T. Rivlin and R. Sibner, *The degree of approximation of certain functions of two variables by a sum of functions of one variable*, Amer. Math. Monthly **72** (1965), 1101–1103.
117. V. A. Rokhlin and D. B. Fuks, *Beginner's course in topology: geometric chapters*, "Nauka", Moscow, 1977; English transl, Springer-Verlag, Berlin, 1984.
118. W. Rudin, *Functional analysis*, McGraw-Hill, New York, 1973.
118a. H. S. Shapiro, *Some theorems on Čebyšev approximation.* II, J. Math. Anal. Appl. **17** (1967), 262–265.
119. M. R. Shura-Bura, *Approximation of functions of several variables by functions depending on one variable*, Vychisl. Mat. **2** (1957), 3–19. (Russian)
120. I. Singer, *Best approximation in normed linear spaces by elements of linear subspaces*, Springer-Verlag, Berlin, 1970.
120a. L. G. Schnirelman, *On some properties of closed curves*, Uspekhi Mat. Nauk **10** (1944), 34–44. (Russian)
121. D. Sprecher, *On the structure of continuous functions of several variables*, Trans. Amer. Math. Soc. **115** (1965), 340–355.
122. ———, *A representation theorem for continuous functions of several variables*, Proc. Amer. Math. Soc. **16** (1965), 200–203.
123. ———, *On the structure of representations of continuous functions of several variables by finite sums of continuous functions of one variable*, Proc. Amer. Math. Soc. **17** (1966), 98–105.
124. ———, *An improvement in the superposition theorem of Kolmogorov*, J. Math. Anal. Appl. **38** (1972), 208–213.
125. ———, *A survey of solved and unsolved problems of superpositions of functions*, J. Approx. Theory **6** (1972), 123–134.
126. ———, *On similarity in functions of several variables*, Amer. Math. Monthly **76** (1969), 627–632.
127. ———, *On best approximations in several variables*, J. Reine Angew. Math. **229** (1967), 117–130.
128. ———, *On best approximations of functions of two variables*, Duke Math. J. **35** (1968), 391–397.
129. ———, *On the existence of best approximations and representations in several variables*, J. Reine Angew. Math. **234** (1967), 153–162.
130. ———, *On functional complexity and superpositions of functions*, Real Anal. Exchange **9** (1983/84), 417–431.
131. J. P. Sproston and D. Straus, *Sums of subalgebras of $C(X)$*, J. London Math. Soc. **45** (1992), 265–278.
132. Y. Sternfeld, *Dimension theory and superpositions of continuous functions*, Israel J. Math. **20** (1975), 300–320.
133. ———, *Uniformly separating families of functions*, Israel J. Math. **29** (1978), 61–91.
134. ———, *Superpositions of continuous functions*, J. Approx. Theory **25** (1979), 360–368.
135. ———, *Dimension, superposition of functions and separation of points in compact metric spaces*, Israel J. Math. **50** (1985), 13–53.
136. ———, *Uniform separation of points and measures and representation of sums of algebras*, Israel J. Math. **55** (1986), 350–362.
137. V. M. Tikhomirov, *Kolmogorov's work on ε-entropy of function classes and the superposition of functions*, Uspekhi Mat. Nauk **18** (1963), no. 5, 55–92; English transl., Russian Math. Surveys **18** (1963), no. 5, 51–87.
138. V. N. Trofimov and L. R. Khariton, *On the error in uniform approximation of functions of two variables by sums of functions of one variable*, Izv. Vyssh. Uchebn. Zaved. Mat. **1979**, no. 8, 70–73; English transl., Soviet Math. (Iz. VUZ) **21** (1979), no. 8, 71–74.
139. A. J. Vaindiner, *Approximation of continuous and differentiable functions of several variables by generalized polynomials (finite linear superpositions of functions of fewer variables)*, Dokl. Akad. Nauk SSSR **192** (1970), 483–486; English transl., Soviet Math. Dokl. **11** (1970), 648–652.

140. Y. A. Vainstein and M. A. Kreines, *Sequences of functions of the form* $f[X(x) + Y(y)]$, Uspekhi Mat. Nauk **15** (1960), no. 4, 123–128. (Russian)
141. A. G. Vitushkin, *On Hilbert's thirteenth problem*, Dokl. Akad. Nauk SSSR **95** (1954), 701–704. (Russian)
142. ———, *Theory of transmission and processing of information*, Pergamon Press, New York, 1961.
143. ———, *Representability of functions by superpositions of functions of fewer variables*, Proc. Internat. Congr. Math. (Moscow, 1966), "Mir", Moscow, 1968, pp. 322–328; English transl., Amer. Math. Soc. Transl. (2) **86** (1970), 101–108.
144. ———, *On representation of functions by means of superpositions and related topics*, Enseign. Math. (2) **23** (1977), 255–320.
145. ———, *On the* 13*th problem of Hilbert*, The Hilbert problems (P. S. Aleksandrov, ed.), "Nauka", Moscow, 1969, pp. 159–169; German transl., Ostwalds Klassiker Exakt. Wiss., vol. 252, Akademische Verlag. Geest & Portig, Leipzig, 1971.
146. A. G. Vitushkin and G. M. Khenkin, *Linear superpositions of functions*, Uspekhi Mat. Nauk **22** (1967), no. 1, 77–124; English transl., Russian Math. Surveys **22** (1967), no. 1, 77–125.

Selected Titles in This Series

(Continued from the front of this publication)

123 **M. A. Akivis and B. A. Rosenfeld,** Élie Cartan (1869–1951), 1993

122 **Zhang Guan-Hou,** Theory of entire and meromorphic functions: Deficient and asymptotic values and singular directions, 1993

121 **I. B. Fesenko and S. V. Vostokov,** Local fields and their extensions: A constructive approach, 1993

120 **Takeyuki Hida and Masuyuki Hitsuda,** Gaussian processes, 1993

119 **M. V. Karasev and V. P. Maslov,** Nonlinear Poisson brackets. Geometry and quantization, 1993

118 **Kenkichi Iwasawa,** Algebraic functions, 1993

117 **Boris Zilber,** Uncountably categorical theories, 1993

116 **G. M. Fel′dman,** Arithmetic of probability distributions, and characterization problems on abelian groups, 1993

115 **Nikolai V. Ivanov,** Subgroups of Teichmüller modular groups, 1992

114 **Seizô Itô,** Diffusion equations, 1992

113 **Michail Zhitomirskiĭ,** Typical singularities of differential 1-forms and Pfaffian equations, 1992

112 **S. A. Lomov,** Introduction to the general theory of singular perturbations, 1992

111 **Simon Gindikin,** Tube domains and the Cauchy problem, 1992

110 **B. V. Shabat,** Introduction to complex analysis Part II. Functions of several variables, 1992

109 **Isao Miyadera,** Nonlinear semigroups, 1992

108 **Takeo Yokonuma,** Tensor spaces and exterior algebra, 1992

107 **B. M. Makarov, M. G. Goluzina, A. A. Lodkin, and A. N. Podkorytov,** Selected problems in real analysis, 1992

106 **G.-C. Wen,** Conformal mappings and boundary value problems, 1992

105 **D. R. Yafaev,** Mathematical scattering theory: General theory, 1992

104 **R. L. Dobrushin, R. Kotecký, and S. Shlosman,** Wulff construction: A global shape from local interaction, 1992

103 **A. K. Tsikh,** Multidimensional residues and their applications, 1992

102 **A. M. Il′in,** Matching of asymptotic expansions of solutions of boundary value problems, 1992

101 **Zhang Zhi-fen, Ding Tong-ren, Huang Wen-zao, and Dong Zhen-xi,** Qualitative theory of differential equations, 1992

100 **V. L. Popov,** Groups, generators, syzygies, and orbits in invariant theory, 1992

99 **Norio Shimakura,** Partial differential operators of elliptic type, 1992

98 **V. A. Vassiliev,** Complements of discriminants of smooth maps: Topology and applications, 1992 (revised edition, 1994)

97 **Itiro Tamura,** Topology of foliations: An introduction, 1992

96 **A. I. Markushevich,** Introduction to the classical theory of Abelian functions, 1992

95 **Guangchang Dong,** Nonlinear partial differential equations of second order, 1991

94 **Yu. S. Il′yashenko,** Finiteness theorems for limit cycles, 1991

93 **A. T. Fomenko and A. A. Tuzhilin,** Elements of the geometry and topology of minimal surfaces in three-dimensional space, 1991

92 **E. M. Nikishin and V. N. Sorokin,** Rational approximations and orthogonality, 1991

91 **Mamoru Mimura and Hirosi Toda,** Topology of Lie groups, I and II, 1991

(See the AMS catalog for earlier titles)